PRACTICAL
LEATHER
TECHNOLOGY

FOURTH EDITION

THOMAS C. THORSTENSEN

TSG Consulting Group

KRIEGER PUBLISHING COMPANY
MALABAR, FLORIDA
1993

Fourth Edition 1993

Printed and Published by
KRIEGER PUBLISHING COMPANY
KRIEGER DRIVE
MALABAR, FLORIDA 32950

Copyright (c) 1984 (new material) by
Robert E. Krieger Publishing Company, Inc.
Copyright (c) 1993 (new material) by
Krieger Publishing Company

Library of Congress Cataloging-in-Publication Data

Thorstensen, Thomas C., 1919–
    Practical leather technology / Thomas C. Thorstensen. — 4th ed.
      p.   cm.
    Includes index.
    ISBN 0-89464-689-3
    1. Leather. 2. Tanning. I. Title.
  TS965.T56   1992
  675'.2—dc20   91-39162
                                                                      CIP

10 9 8 7 6 5 4 3 2

*Dedicated to the
international, technical, economic, and
environmental development of the leather industry.*

ACKNOWLEDGMENT

In the revision of the chapter on finishing, Mortimer Greif was most helpful in supplying information that would only be known to a person experienced in the development of new finish systems. His help and suggestions are gratefully appreciated.

CONTENTS

PREFACE TO THE FOURTH EDITION

In preparing the fourth edition of this book it came to me as a surprise that it has been over twenty years since I wrote the first edition. What is even more surprising is that much of the book still stands as an explanation of some of the fundemental technology of the industry.

Many of the photographs were taken for the original edition and haven't been changed. In many tanneries the same machines are still in use. The new machines are larger and more accurate, but, in many cases the same mechanisms are employed. Where a picture shows the operation of the machine, the original photographs have been kept. The primitive tanning method pictures have been kept for general interest.

In rereading the prefaces for the first, second, and third editions of the book, the changes in the industry were highlighted. In considering these changes at this time the continuation of their direction is evident. The factors that force change in the industry are still the industrialization of developing nations, the environmental aspects of the industry, and the continuing search for quality of the product and efficiency of production. The environmental impacts of the industry can no longer be ignored. Extensive pollution control projects are now a part of the operation wherever leather is being made. The search for quality and efficiency has resulted in the same machinery and the same advanced technology being used in tanneries the world over.

The changes in our society in fast, accurate communication and international travel have overcome many of the problems of doing business on an intercontinental basis. The tanneries in the developing nations are now producing quality leather goods that are accepted in the international markets.

This edition includes information gathered from friends in the industry on five continents. I salute them all for their skill and artistry. The exchange of information in publications and on an informal basis is a part of the tanner's guild. We are friendly competitors who recognize quality when we see it whether it is in technology or in people.

PREFACE TO THE THIRD EDITION

The original edition of this book in 1969 covered the industry as I saw it at that time. The second edition in 1976 included changes that were evident due to the technological economic developments on a worldwide basis. Now we are again given a chance to consider where we are and where we are likely to go. At the time of this revision, we are concerned with the generally poor economic situation, inflation, and unemployment. In the United States, there is a concern over what appears to be, to the American tanners, unfair regulation, import duties, and export subsidies in the international market.

The world pattern of leather manufacturers has changed. There has been a very significant development of the industry in Argentina, Brazil, India, and other nations, primarily geared to the export of leather and leather goods rather than raw materials.

Pressures for environmental responsibility have also resulted in great changes. In the United States, only those companies that could muster the funds and the desire to continue have survived. The environmental problems will continue to affect the industry, but in my opinion, the United States has met the challenge.

There has been great fluctuation in hide prices, particularly on the international market. This also has taken its toll. When hide prices are on the rise, tanners must honor previous sales agreements and sell at less than replacement costs. When hide prices fall, the customer forces lower prices, and losses again occur. Changes in inventory value have placed many companies in a precarious position.

As a result of these problems, we have seen many companies close their doors, and many of the obsolete and marginal plants have been eliminated. Stronger, more financially secure companies have taken over their competitors and grown into fewer and larger multiplant operations. Some of these are part of large conglomerates.

The leather industry in the United States has matured. Tanneries are no longer run like one-man bands. Responsible decisions are made in technical confidence in a competitive world market, by people with managerial skills and social responsibility.

Along with the changes that have taken place in the American and European leather industries, we have observed also tremendous changes taking place in the so-called underdeveloped nations. Historically, Asia, Africa, and South America exported their raw material to Europe and the United States where the hides would be made into leather and sold as quality fashion items. The underdeveloped countries have raw materials and a willing labor supply capable of operating an industry in a labor intensive manner with very little machinery. Manufacture of leather by such primitive methods as are shown in some of the photos in this book was common in these areas for centuries. Some of this primitive manufacture is still going on.

Intensive efforts have been made to develop a viable leather industry to export leather and if possible leather goods to North America and Europe. This is of course an effort to improve the employment of the people and to export goods for hard currency. To do this the primitive tannages, although perhaps quaint in their craftmanship, are an economic loss. Instead, the valuable raw materials must be manufactured into saleable goods which can find a broad ready market. The modernization, therefore, of the leather industries in the emerging nations does not go through a gradual stepwise change repeating the changes that were made in the industrialized nations decades ago. The technology steps immediately to modern production. Developments such as the most modern accurate shaving machines, splitting machines, and most modern drying equipment are adopted very rapidly. These developments are to ensure quality of the product for unquestioned acceptance in the world market. The updating is not limited to machines, but is evident in the fashion and modern commercial practices. These countries have extensive programs for training people, obtaining new equipment, and assuring that the factories so developed have an added economic incentive for their existence.

We would expect to see, during the useful life of this book, continued expansion of the leather industry in the emerging nations.

Thomas C. Thorstensen

PREFACE TO THE SECOND EDITION

Since the original publishing of the book in 1969 there have been significant changes in the leather industry and its method of manufacture. Changes in technology as well as the effects of pollution regulations will be incorporated in the second edition. The main factors contributing to these changes are:

1. The technological development of the emerging nation.
2. Environmental problems.
3. The development of new specialty chemicals.
4. The increased cost of labor.
5. The world shortage of leather making materials.

In re-reading the book some five years after its original writing I find there are many areas that I would like to expand to cover information in greater detail. The generalized approach necessary to cover a broad subject forces some omission of very significant material. With the idea of keeping the book simple and readable for those who are on the fringes of the industry or need a basic starting point, it is deliberately kept simple and we are not attempting to cover detailed information on all phases of production or the science involved in leather technology. The path between oversimplification with inaccuracies and confusing details is narrow indeed.

Thomas C. Thorstensen

PREFACE TO THE FIRST EDITION

It is the aim of this book to present a general introduction to the principles of leather technology, including consideration of economic factors and leather usage. Being an introduction to the field, it is designed to be of use primarily to persons new to the industry or working in the allied fields. A descriptive and qualitative approach is used wherever possible. Within the limitations of space and scope, the book is not a rigorous review of current literature nor is it a detailed compendium of commercial formulation.

It is intended that the book be read in its entirety so that the reader can appreciate the interrelationship of processes. When the book is used as a reference, the entire chapter should be read, rather than just a single paragraph or sentence to cover the question of the moment.

In each chapter, where applicable, a section called "Looking Ahead" is included. In this section the possible practical application of current research is considered. It is hoped that these opinions will stimulate the reader's thinking and emphasize the changing patterns of the industry.

Many people have helped in the preparation of this book. Tanners, machinery manufacturers, chemical suppliers, and government agencies have been very helpful in supplying photographs, technical data, and other pertinent information. In this regard we are particularly indebted to William Roddy (Tanners Council Laboratory) for the collection of photographs of skins, leather, and grain; Dr. Joseph Naghski of the Department of Agriculture was very helpful in supplying additional information. Some specific chapters were reviewed for technical accuracy by people versed in the different fields, and their help is hereby acknowledged: E. B. Thorstensen, D. F. Lord, H. R. Bennett, T. Ohsugi, Edward Kaine, Arthur F. Schroeder, John J. Riley, and many others.

The drawing and art work are by Peter Hynes of The Massachusetts Institute of Technology. The author would also like to thank Mrs. Phyllis Kean and Miss Beverly Banks for their work in the manuscript preparation. The editorial assistance of Mr. G. G. Hawley is also greatly appreciated.

Thomas C. Thorstensen

HIDES AND SKINS— COMMERCIAL ASPECTS

1

The leather industry may be regarded as a bridge between production of the hide as a by-product of the food industry and its manufacture into shoes and wearing apparel, for which it provides a basic raw material. The technologies and skills involved in the production of meat and those required in the production of usable goods from leather are widely different. The two groups of people having these different skills have very little in common. This separation of skills and degree of specialization has existed in all but the most primitive societies.

The production of leather is a long and complicated process, and certainly not one which can be embarked upon successfully without specialized skills. The hide is in immediate need of some form of preservation from decay. Once this is accomplished it can be shipped great distances or stored until used. The cured hide, regardless of source, is an article of international commerce. The demand for a particular type of hide may have no relation to the supply. The amount and type of meat in the diet of the people determines the supply of hides. The demand for shoes and other leather goods is independent of the supply of hides. For example, since the appetite for beef exceeds the need for leather goods, the United States is a hide-exporting nation. Japan, however, has a strong demand for leather goods but a limited hide supply; Japan is a hide-importing nation.

An additional factor in the supply of hides is the type and quality of skins being produced. Some parts of the world may produce light skins from the slaughter of sheep and goats, and yet have a demand for heavy hides. It is not uncommon, for example, for sheepskins to be produced in New Zealand, converted into leather in the United States, and made into garments in Europe. Also, skins partially processed in India may be further tanned in Europe and sold as finished goods on the American market.

Any hide or skin, therefore, is a valuable article of international commerce and must be handled to the best economic advantage. The decision as to how the skin is to be cured and marketed is affected by the ultimate user of the hide, but other factors also influence the curing. The availability of curing materials, the weather, local labor costs, local sociological and religious customs, and business customs all play a part in the technological and commercial system of tanning.

1

AMERICAN BEEF CATTLEHIDES

One of the most important factors in the leather industry of the world is the large production of cattle hides in the United States. With a high standard of living and an appetite for beef, Americans presently consume more than 35,000,000 cattle a year. Most of this is beef from the Midwest. The quality of beef, when sold for slaughter, is judged on the basis of contour of the carcass and the fiber structure of the meat. One of the most significant factors is the presence of a marble effect, due to fat, in the loin section of the carcass.

The Western beef steer is generally born in a grassland area of the Southeast or Southwest. A young calf is allowed to grow on grass for approximately one year, at which time, at a weight of about 400–500 pounds, it is shipped to the grain belt area of the Midwest, put in a feeder lot, and fed a high protein diet of soybeans, corn and other grains to bring it up to market weight. At market time these Western steer beef animals generally weigh about 1100 pounds and are between two and three years old. They are mostly Hereford or Hereford-type, with Black Angus second in number. When the steers are ready for market they are taken to a stockyard to be sold by competitive bid to various packers.

The packing industry operates on a very fast turnover of capital and a low profit margin based on sales. The packer buys an animal for cash and sells the meat for cash. The price of the meat is about equal to the price of the animal, and expenses and profit must come from the sale of by-products.

The packer may have considerable capital tied up in hide inventory because the curing of the hides takes approximately a month in a salt pack. The value of the hide may be between 5 and 10% of the value of the meat, depending on prices and classifications. A further risk during this long curing time is the danger of loss due to market price changes.

In a modern American packing plant, the packer buys the cattle on competitive bid and slaughters them immediately. Cattle that are to be slaughtered in the morning are usually purchased late in the afternoon, the day before, so that storage costs are only necessary for one evening at the stockyard. The next morning, cattle will be purchased on competitive bid for slaughter during the day. The pricing is very closely controlled by an elaborate communication system from stockyard to stockyard involving Telex and two-way radios to the buyers in the yard. A price change of twenty-five cents per hundred weight on the cattle in Omaha, for example, will be reflected in the prices at other stockyards in a matter of a few minutes. Cattle are purchased against meat orders and the sides of beef are shipped out usually within forty-eight hours. The amount of capital tied up by the packer in meat inventory is, therefore, minimal. Under the conventional hide cure, the value of inventory of hides exceeds the value of inventory of meat. Hides, therefore, are sold for cash and are usually paid for with a sight draft bill of lading. The desire for reduction in inventory has pushed for more effective curing and hide processing methods and will be outlined in further chapters of this book.

AMERICAN DAIRY CATTLEHIDES

Dairy cattle are bred and raised strictly for their producton of milk; a dairy cow is marketed for slaughter only when it is no longer an efficient milk producer. Dairy cattle are found more in the grazing areas than in the grain areas. The hides from these cattle are thin and spready, and contain less fat than the beef cattle types. They are generally slaughtered by smaller packing companies.

Calfskins

Calfskins are a by-product of the dairy industry and are available in dairy producing areas of the United States and Europe. The skin is highly valued because of the fine grain structure and usually commands three to five times

TABLE 1

U.S. Commercial Slaughter of Cattle

Year	Cows and Heifers	Steers	Bulls	Total * Inspected	Total
		Millions of Head			
1976	20.8	17.3	.9	39.0	42.6
1979	14.7	16.3	.6	31.5	33.7
1982	16.7	16.3	.8	33.9	35.8
1985	17.8	16.2	.7	34.7	36.3
1988	16.5	16.8	.6	34.0	35.1

*Federally Inspected Slaughter

the price of heavy cattle hides. Calfskins, therefore, are cured with great care, being washed and thoroughly cured with fine-grain, pure salt.

Dairy cattle are bred for milk and are raised in rich pasture lands as are found in Wisconsin and New York. Since calfskins are a by-product of the dairy industry, they are a relatively small source of leather-making raw materials in the United States.

In Europe, there is a preference for veal. There is also a large dairy industry. As a result, there is an excellent calfskin supply in France and Germany.

(a)

(b)

Figure 1a, 1b Hide and skin defects.* **(a)** Inspection of pickled skins. The use of a light to shine through the skin permits quick inspection for defects and fiber structure in sheepskins. (*Courtesy L. H. Hamel Leather Co.*) **(b)** Inspection of dried goatskins prior to soak. (*Courtesy L. H. Hamel Leather Co.*)

* The subject of hide and skin defects has been reviewed extensively. Tancous, J. J., Roddy, W. T., and O'Flaherty, "Skin Hide and Leather Defects," Cincinnati, Western Hill Publishing Co., 1959.

Goatskins

Goats are hardy animals that can live on a wide variety of foods and can supply meat and milk. They are adaptable to difficult climates and are popular in Asia, Africa, and South America. The original sources of many of the goatskins are villages of widely diversified areas, so the quality varies greatly. Also important in determining the quality of the goatskins are the type of animal, the method of slaughter, the method of cure, and the marketing practices of the area of origin. Between the villages and the world market there is a system of collectors and dealers. The practices of handling skins and the business methods of each area have long been established by custom and tradition. These factors, different in each part of the world, are important in determining the quality, characteristics, and price of the skins. The skins are identified by the area of origin and are sold either on a size specification by the dozen or by the pound.

Sheepskins

Pickled, salted, and dried sheepskins are available from many parts of the world, notably South America, Africa, Australia, and New Zealand. Probably the largest single factor is the sheepskin industry of New Zealand, where there are approximately two hundred locker plants where the sheep are slaughtered and skins processed. As the value of the wool exceeds that of the skin, it is customary to remove the wool from the skin by woolpulling methods (described fully in Chapter 6), and the skins are sold in the pickled state. Because of the long wool and the high fat content of the skin, curing of these skins in a salt pack is rather difficult. The slaughter of sheep and lambs in New Zealand follows the seasons and conditions of pasture, creating a difficult purchasing situation for the tanner.

The slaughter of sheep in the United States does not follow this seasonal cycle as closely. Sheepskins are generally salted. Domestic sheepskins are salted and sold for use in grain leathers, suedes, or shearlings. Since much of the weight of the sheepskin is made up in the wool and the grease associated with the wool, sheepskins are sold by the piece.

Pigskins

Pigskins have a unique structure in that they are very greasy and the hair penetrates through the usable skin, leaving holes in the leather after it has been processed. There are three general types of pigskin coming from three widely different sources.

(1) Wild Boar: The wild boar, or peccary, is a very lean animal and produces a very tough grain-type leather suitable for gloves, garments, and shoes.

(2) European Pig: In Europe, particularly in eastern Europe, it is customary to skin the pigs and salt the skins. Salted pigskins can then be processed as hides to make a wide variety of grain leathers. This skin is less fatty than the pig of the United States, and, in general, is more desired for its leather-making qualities. Polish and eastern European pigskins are exported to the United States in relatively large quantities.

(3) United States Pig: These pigs are usually scalded when slaughtered. The hot water loosens the hair and it is brushed from the skin. The hair is recovered, but the skin remains with the carcass. Later, when the pig is cut into hams, loins and bacon, the ham and bacon parts will retain a portion of the skin, but the loin area will be recovered as pigskin. These pigskins are used primarily for gelatin rather than for leather.

Recently a system has been developed whereby the best portions of the domestic pigskin can be recovered in a condition suitable for suede shoe leather. This system, along with an excellent marketing program controlled by one company, has resulted in wide American acceptance of pigskin suede shoes.

With the shortage of leather-making materials on a worldwide basis, much attention has been placed on the possibility of improving or increasing the use of pigskin, particularly American pigskins for a leather-making raw material. It was anticipated in the late 1970s that shortly (early 1980s) almost all of the pigskin slaughter in the United States would be converted to a curing system whereby more leather could be made. With the adverse conditions in the leather industry, particularly in the United States, this anticipated demand for pigskin in the United States as a leather-making raw material will depend upon the demand of the pigskin for edible use such as gelatin manufacture and the total economics of the world leather industry.

Reptiles

There are three classifications of reptiles used for commercial leather production: snake, lizard, and crocodile or alligator.

Snakeskins from the python, boa, watersnake, and cobra have the most significant commercial value. The snakeskins are obtained from the animals in the wild state and are salted or tacked to dry.

Lizard skins are treated in a similar manner to snakeskins.

Alligators and crocodiles may come from South America, or from the islands of Southeast Asia. Smaller alligators and crocodiles are cut down the belly, and the back (called hornback) is used. The larger alligators and crocodiles are cut down the back, and the square belly pattern is used. Most crocodiles and alligators used in commercial leather manufacture are caught wild. Prices of these skins depend upon the size, based on inches of width. Although a wide skin may be longer, therefore having a larger area of skin available, the grain patterns are not as fine. The inch-of-width measurement is a fair means of determining actual value. The lack of consistent quality, however, and the difficulty of obtaining a reliable source of supply make the reptile skin business very speculative.

The hunting of many reptiles has greatly decreased because many species have become endangered. Alligators and crocodiles are protected in most of their natural habitat as are some snakes. In the case of alligators and crocodiles, protection of the animals in their natural environment and raising them on "farms" periodically produces an excess population. Permitting "controlled" hunting results in the harvesting of some of the skins. The production of many types of reptile skins follows the same pattern.

Specialty leathers

There are, in addition to the hides and skins listed above, a number of other categories of specialty leathers which have become popular at one time or another. These include ostrich, sharkskin, elephant, kangaroo, turtle feet, and even frogskins. Of these, sharkskin and kangaroo are available on the commercial market in greatest quantities. In addition to a distinctive grain pattern, sharkskin is very tough and is used where high abrasion resistance is desired. Kangaroo is a very tight-grained leather, very soft and similar to kid. It is used as a shoe upper leather and is also popular in athletic shoes.

INTERNATIONAL TRADE IN HIDES AND SKINS

Hides and skins, as by-products of the food industry, are found in the meat-producing areas. The demand for leather is not the same as the demand for meat as food. In India, for example, the cattle population is very large and is a vital part of the economy. The demand for meat is very low due to religious and cultural customs of the people. In New Zealand, Argentina and many other countries, sheep or cattle are raised for a major export of meat. Such countries are exporters of hides and skins. The United States is a producer of over forty million cattlehides each year, and of these, over half are exported.

The international market in hides and skins is constantly shifting in accordance with the demand for leather in various parts of the world. The demand for leather is increasing as the standard of living in the developing nations rises. The world leather production and hide shipments are in constantly changing patterns.

TABLE 2

Key U.S. Hide Imports, Exports and Production in Millions*

Year	Total Slaughter	Exports	Imports	Net Exports	Leather Produced
1976	43.5	25.3	1.0	24.3	20.2
1979	34.4	23.7	.6	23.0	15.0
1982	36.6	23.2	.7	22.5	15.0
1985	38.2	25.4	1.0	24.4	12.6
1988	36.8	25.1	.6	24.5	11.4

*From Leather Industries of America Inc.

TABLE 3

U.S. International Trade in Leather and Leather Goods

	$ U.S. Millions			
	Leather		Shoes and Leather Products	
	Imports	Exports	Imports	Exports
1976	181	139	2,171	111
1979	284	250	3,564	189
1982	318	279	4,476	237
1985	394	284	7,626	205
1988	748	506	11,001	379

United States Leather Exports are: Upholstery, Wet Blues, Crust and Finished Leather.
United States Leather Imports are: Finished Leather, Upholstery, Garment, Glove and specialty leathers.
United States Shoe and leather products Imports are: Shoes (65–70%), Garments (7%), Handbags (7%).

COMMERCIAL PRACTICE OF THE AMERICAN HIDE INDUSTRY

Cattlehides have been traded on the Commodity Exchange. This practice was helpful to the tanner because it was possible to protect the value of the inventory in production by the purchase of hide futures. The production of leather is no longer a several-month process. With the development of modern high speed tannages and the increased importance of the international market, the sale of hide futures is no longer a significant part of the American leather industry. The rules used by the Commodity Exchange give good information on the nature of the business and as such are included in our discussion of the practices of the industry.

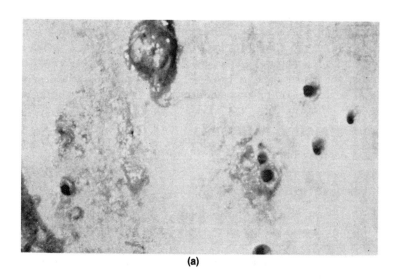

(a)

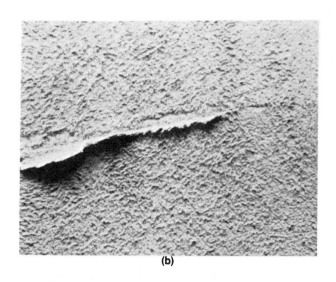

(b)

Figure 2a, 2b Hide and skin defects. **(a)** This photograph shows a grub. It is a parasite in the larva stage that lives in cattle and causes holes in the bend. (*Courtesy Tanners Council Laboratory*) **(b)** Score as seen in the flesh side of the leather. This type of defect is caused in the flaying of the skin. (*Courtesy Tanners Council Laboratory*)

Hides are sold for cash. The packer does not allow even a few days' credit. To speed payment, a sight draft bill of lading may be used, and the tanner may pay for the hides before he gets them and, in most cases, before he sees them. In order to assure the tanner of the quality of hides he is buying, negotiations seldom take place between the packer and tanner. Instead, the tanner employs a broker to represent his interests. The broker provides an inspector at the time of "take-up" to check grades, weights, hide defects, and tare (manure) allowance. The hide broker receives a fee from the tanner for his services. A broker is not the same as a dealer; a dealer buys and sells hides on his own account and is a speculator in the market.

Factors affecting hide value

Hides are described by type, weight classification, and month of take-off.

Seasonal variation

Under the Commodity Exchange rules, a discount is allowed depending upon the month of take-off. Under negotiated purchases of specific lots of hides this does not apply, but the Commodity Exchange indicates normal season value fluctuations. Using the summer months as par, the discount increases up to 6% during January, February, and March. The decrease in value during the winter months is due to the larger amount of hair, fat, and other general changes in hide condition due to the cold weather adjustment of the animal.

Seasonal Price Discount.

Jan., Feb., Mar.	6%
Apr.	4%
May	3%
June	1%
July, Aug., Sept.	No discount
Oct.	1%
Nov.	3%
Dec.	4%

The value of hides changes with the characteristics of the hides and with the seasons. Summer hides have less fat and hair (in temperate climates) than winter hides. As a result, if the hides are sold from storage the month of take off is important and is specified in the contract of sale.

From these data we can make a few calculations to illustrate actual hide values. Assuming a standard 40,000 pound future contract at twenty cents per pound for 53–54 pound average hides, the number of hides in the shipment, the cost of the hides, and the price per hide can be calculated. From these figures it can be seen that the price per hide undergoes only a small change, ± 10%, over a wide range of weights. This is due to the compensating factors of higher weight and lower price per pound.

The value of a hide is determined by the value of the leather than can be made from it. Since leather is sold by the foot, small hides (being thinner) are more valuable, as they give more area per pound of hide substance purchased. A second factor, even more important, is that smaller hides have finer grain, so their leathers will command a higher price. Although market prices on hides may fluctuate widely over a short period of time, the relative value of one type of hide as compared with another can be changed only by different patterns in leather types. Hides, being a versatile raw material, lend themselves to the tanner's skill to make a wide variety of leathers, and thus the fashion factor becomes less significant and the relative price relationship remains.

Weight variation

Under Commodity Exchange rules, the hide futures are sold as 53–54 pound average for domestic green, salted, unfleshed hides. It is impossible, of course, to predict the exact actual weight average of the hides, so adjustment must be made. A discount is used for weight averages above the 53–54 pound norm and a premium is paid for weight averages below the norm. A few values are taken as an example:

TABLE 4

Value of hides based on weight discount allowances 53–54 average 40,000 pounds standard shipment at 20 cents per pound.

Average wt. (lb)	Premium or discount	Shipment value	Price cents per lb.	No. of hides	Price per hide
41 lb	19½% Prem.	$9560	23.90	975	$ 9.84
45 lb	12 % Prem.	8960	22.40	888	10.08
50 lb	4½% Prem.	8360	20.90	800	10.48
53–54 lb	Base	8000	20.00	755	10.62
60 lb	7 % Disc.	7440	18.60	667	11.14
65 lb	12 % Disc.	7040	17.60	615	11.46
70 lb	17 % Disc.	6640	16.60	571	11.66
77 lb	24 % Disc.	6080	15.20	519	11.74
85 lb	32 % Disc.	5440	13.60	470	11.58

The "modern trim"

The selling price of a hide on a pound basis is less than the selling price of the meat on the carcass and considerably more than the price of *offal*.* It would be to the packer's advantage, therefore, to leave as little flesh on the hide as possible, provided this flesh remains on the carcass. The offal, on the other hand, is more valuable as hide than as rendering stock. As a result of these factors it was the packing industry practice for many years to leave the ears, tail, shanks, snouts, and other appendages on the hides. In the case of collector hides (farmer slaughter), it was common for a tanner to find even more weight-adding components such as hooves, horns, and an occasional surplus piece of farm machinery. The added weight was carried through the cure and shipped with the hide. The tanner then trimmed off these parts before starting his process.

The economic loss from the old trim practice is obvious. As a result of the cooperation between the farmers, packers, dealers, and brokers through their respective associations, O'Flaherty and Roddy of Tanners Council were able to define a new trim system which has now become the standard of the industry.

The modern hide trim results in a higher price per pound for the tanner but a net economic gain in the saving of freight. The packer gains more offal for his rendering plant. The modern trim is being taken a step further

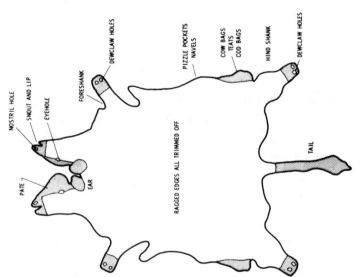

Figure 3   Modern trim diagram, accepted as standard by the American industry.

* The term *offal* as used here refers to the inedible portions of the carcass (head, trimmings, entrails, etc.) that are rendered for fats by the packer.

by some hide dealers in that hides are being segmented for the removal of double shoulders and/or bellies so that they can be processed into different types of leather.

Fleshing of hides

In the mid 1950's the practice of prefleshing hides came into extensive use. The hide, as removed from the animal, has about 20% of its weight as excess flesh. The removal of this flesh will allow better cure and give a lighter shipping weight. Prefleshing also saves one step in the production process for the tanner. The benefits of this system work both ways. The tanner saves in freight and processing costs but loses some offal yield (flesh). The packer gets the offal and a better cure.

In calculating the relationship between fleshed and unfleshed hide in accordance with Commodity Exchange classifications, the assumption is made that the flesh weight, properly trimmed, is 22% less than unfleshed hides.

Hide classification

The terms used to describe the classification of hides are based on the conditions of the industry half a century ago. Terms such as "Texas steers,"

TABLE 5

Seasonal Price Variations of U.S. Hides Cents/Pound

Year	1976	1979	1982	1986	1989
Month					
Jan	30.9	76.0	59.0	58.8	90.0
Feb	30.5	88.0	58.0	55.0	95.0
Mar	38.1	110.4	54.7	58.3	100.0
Apr.	42.2	111.8	57.4	60.0	100.0
May	38.6	108.0	51.3	57.6	92.0
Jun	35.8	90.6	48.2	55.7	82.8
Jul	37.8	81.4	44.8	56.0	80.3
Aug	39.2	79.6	44.5	56.0	80.3
Sept	39.7	74.8	43.9	58.0	86.0
Oct	34.3	74.5	40.0	59.0	86.0
Nov	34.1	73.1	43.3	60.5	82.7
Dec	39.5	77.8	48.0	60.0	79.2
Average	37.1	87.2	49.4	58.0	88.6

"Colorado steers," and "big packer hides," for instance, no longer mean hides from these sources, but rather describe types. See Table 3.

Prefleshed brined hides. This refers to hides that are prepared by prefleshing, modern trim and prefleshing, and brining as described in the previous section. Many tanners work only prefleshed hides for a matter of efficiency. One disadvantage to the prefleshed brine cured hides is that the total salt content is usually somewhat larger than by the pack method, so the keeping power of the hides in long storage may not be as good.

Big packer hides. This term originally referred to those hides obtained from one of the "Big Four" meat packing firms. A generation or so ago these packers operated very large packing plants in the Midwest, particularly in the Chicago area. These plants had good take-off techniques and good curing systems, for that time. As a result, the big packer hides were known to have less butcher cuts and were well cured. Now, with the advent of efficient smaller packers capable of producing hides of "big packer" quality, the term refers to hide quality rather than to size of the packing plant.

Small packer hides. This term refers to hides of better quality than country hides but not of sufficient quality to meet big packer standards. These hides generally have some butcher cuts and a slightly poorer quality cure than big packer hides.

Collector hides. These are hides and skins which are obtained by dealers from retail butcher shops, locker plants, and private farm kill. They are generally of poor quality, rather badly damaged by butcher cuts, and quite often poorly cured. A similar classification is that of "country hide."

All weight hides. This term is applied to shipments of hides which contain a spread in weights not defined in one of the specific commercial weight categories. It usually means the purchase of a shipment of hides from a particular packing plant or collector, the hides of which have not been graded in accordance with weight classification.

Frigorifico hides. These are hides from South America, particularly Argentina, Uruguay, and Brazil, which are cured by the Frigorifico method. In this cure, the skins are washed in brine and then placed in salt pack.

Colorado steers. This term originally referred to the branding methods used in Colorado and now refers to hides with large side brands. Colorado steers are sorted by weight, 58 pounds and up as "heavy," 48–58 pounds as "light." They are used primarily for the manufacture of sole leather.

Texas steers. Texas hides referred originally to hides from the Texas area and now refers to hides of the general Texas description. These hides are of small pattern and plump nature. They are generally desirable for shoe upper leather.

Kosher hides. The term refers to hides from animals killed by having the throat cut according to rabbinical law. This cut is evident in the pattern. In modern practice the heads of Kosher hides are usually removed.

TABLE 6

Domestic Packer Hides.[a]

Grade	Description
Side branded hide	A hide being branded on either side and/or both sides, forward of the break in the flank. Exception: Cheek brands.
Butt branded hide	A hide being branded on either butt area and/or both butt areas, back of the break in the flank.

Native steers (free of brands)	Unbranded steer hides.
	  heavy and light
	    ex-light

Butt branded steers	A butt branded steer hide. heavy and light
Colorado or side branded steers	Side branded steer hides. heavy and light
Heavy Texas steers	A side branded steer hide, but of narrow, close compact pattern and plump. heavy
Light Texas steers	Same description as that of heavy Texas with the exception of the weight. light
Ex-light Texas steers	Same description as that of heavy Texas with the exception of the weight. ex-light
Heavy native cows (free of brands)	Unbranded cow hides. heavy
Light native cows (free of brands)	Unbranded cow hides. light
Branded cows	A branded cow hide. all weights May include ex-light. Branded steers.

[a]As defined in standard practice by the Tanners Hide Bureau.

15

Slunk. This is an unborn calf obtained at the time of slaughter of the cow. Slunk skins make very fine suede leathers and command exceptionally high prices.

REFERENCES

"Dictionary of Leather Terminology." New York, Tanners Council of America, 1957.

Federal Code of Regulations, Animal Health Division of Memorandum, 593.5, part 95.24.

"Hide and Leather and Shoes Encyclopedia of the Shoe and Leather Industry," Chicago, Hide and Leather Publishing Co., 1941.

McLaughlin, G. D., and Theis, E. R., "The Chemistry of Leather Manufacture," ACS Monograph 101, New York, Reinhold, 1945.

"New Zealand Pelts," Massachusetts, Cavanaugh, G. P., and New Zealand, Mair and Company, Ltd., 1960.

Stiasny, E., "Gerbereichemie," Steinkopff, 1931.

Tancous, J. J., Roddy, W. T., and O'Flaherty, F., "Skin Hide and Leather Defects," Cincinnati, Western Hills Publishing Co., 1959.

U.S. Department of Agriculture, "Finishing Beef Cattle," Farmers Bulletin No. 2196, 1964

U.S. Department of Agriculture, "Industrial Methods—Seven Facilities for Cattle Slaughtering," Market Research Report 436, Plants in the Southwest, 1961.

U.S. Department of Agriculture, "Livestock Slaughter," 1966-1967.

U.S. Department of Agriculture, "Price Spreads for Beef," Miscellaneous Publication No. 992, 1965.

U.S. Department of Commerce, Commercial Standard CS 268, Modern Trim, O'Flaherty, 1965.

Wilson, J. A., "The Chemistry of Leather Manufacture," Vol. 2, New York, Reinhold, 1929.

HIDE AND SKIN STRUCTURE

2

A living animal can be considered to be a separate organism from its skin coating. If we pick up a cat, we find that the skin moves very easily over the surface of the body and, in essence, it has the appearance that the cat is living inside a fur bag. This appearance is true, to a greater or lesser extent, of almost all animals. The skin, however, is not only a simple outside protective covering, but it has other vital physiological functions as well. These functions include maintenance of body temperature, the excretion of body wastes, protection against physical and bacterial damage, etc. Of the major functions of the skin, that of protection is the most significant with regard to the physical characteristics of the finished leather. Some animals are protected primarily by a heavy coating of hair, scales, or feathers, and others by the denseness of the network of fibers of the skin itself.

COMPARISON OF HIDES AND SKINS

By considering the functions of parts of the skin and the living habits of animals, we can observe structural relationships. From this information we can anticipate the leather-making potential of any animal. Much knowledge of the nature of skin fucntion has been provided by microbiologists and histologists who have investigated the anatomy of skin, both human and animal. Figure 1 is a cross section of cattle hide from which we can see the major portions of the skin and consider their functions.

The predominant feature in Figure 1 is the hair follicle and the growth of the hair from it. The hair itself has a large bulb-shaped root with a small cup at the bottom. During the growth of the hair the follicle is fed through a small blood vessel, and the protein and other materials in the hair follicle form into cells in the hair root. As the hair grows, the cells move toward the surface and become more elongated and, by the time the fiber of the hair has reached the surface of the skin, these cells are spread out into long, thin structures within the hair. The outer surface of the hair consists of protein materials which gradually harden as the hair grows, so that upon emergence from the surface of the skin through the hair follicle, the outer layers have developed a hard, scaly appearance. Hair is primarily keratin, a sulfur-bearing protein.

The lining of the hair follicle is also made of protein of the keratin type, and the keratin line follows along the surface of the skin down into the hair follicle, and up out again. This extreme outer layer of the skin is called the epidermis, and it is hard, quite inert chemically, and constantly in the state of flaking. The cells from underneath the outer layer of the epidermis are forming and pushing outward, giving constant new growth to protect the outer layer of the skin.

Halfway down the hair follicle is a duct leading from the sebaceous glands. These are oil glands which release oils into the hair and onto the surface of the skin. These glands are more prominent in animals having heavy fur because each hair fiber must be kept lubricated. In addition, the sebaceous glands are vital to the maintenance of proper body temperature in many warm-blooded animals. The sudoriferous or sweat glands release water and undesirable body wastes through the pores of the skin. The evaporation of the moisture from the perspiration causes lowering of the body temperature. In the living organism heat is constantly being generated and this excess heat must be dissipated. Through a proper balance of the rate of excretion between the oil glands and the sweat glands, the quantity of moisture being evaporated is controlled and the body temperature maintained within very close limits.

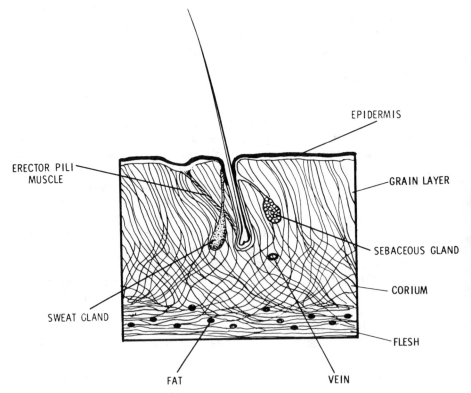

Figure 1   A cross section of cattle hide showing major parts of the skin.

Near the surface of the skin, the fiber structure has a pattern related to the slope of the hair follicles. The fiber structure through this area is generally quite fine in nature. Below the layer of the bottom of the hair root, the structure becomes more random and may have an "angle of weave" in the neighborhood of 45°. The fiber structure here is fairly dense, heavy, and tangled. Closer to the inner layers of the skin the pattern of the fibers more approaches the horizontal, and the longer fibers run parallel to the surface of the skin. The network of collagen fibers, having a fine grain on the surface and full, heavy fiber in the middle, gives the hide its leather-making properties. The shape and texture of these fibers give leather its unique properties of utility and beauty.

In addition to the collagen, there exists in the skin another network of fibers, elastin, which serves to strengthen the hide further. Elastin is an almost chemically inert fiber.

There are a number of muscles related to the skin. The one most generally referred to is the erector pili muscle, which runs down from the opening of the hair follicle at an angle somewhat less steep than that of the follicle. This muscle is used in the animal to make the hair stand on end in times of danger, to give greater protection, and to make the animal look larger. There is also some evidence that the pulling of this muscle will release more oils from the sebaceous glands, and that it is part of the raising of goose pimples when subject to a chill. Below the surface of the skin is another set of muscles running parallel to the surface and extending from near the shoulders backward in a broad fan-type pattern towards the hind legs. The twitching of a horse to shake off flies is a common sight, and it is attributed to this type of muscle.

The muscles in the hides and skins are of relatively little importance in leather manufacture and must be removed, either physically or chemically, during the processing.

The spaces between the collagen bundles are used in many animals as an area for the storage of excess fat, or reserve food supply in the form of fat. Fat depositions in hides are of two broad classifications: those fats used for physiological functions of the skin—hair lubrication and thermostatic control, and those deposited as food storage.

In the spring of the year, when an animal sheds its heavy fur, the hair falls out by the roots and the new hair grows from the same follicle. Such changes result in seasonal differences in the structure of the hair, which may be of important commercial significance in maintaining seasonal quality in leather.

The skin contains a complete system of arteries and veins. Arteries can be distinguished from veins because they are built to withstand higher blood pressure in the animal and are coated on the inside with a light layer of fatty material.

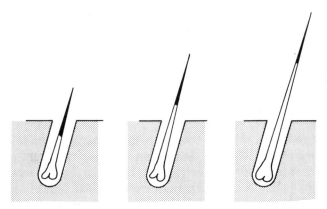

Figure 2 The pattern of hair growth typical in cattlehide and other warm-blood animals. The new hair grows in the follicle with a fully developed root. The new growth of the hair is at the base of the hair follicle.

Comparative structure of hides and skins

Hides and skins differ in their structure, depending upon the habits of life of the animal, season of the year, age, sex, and breeding. In the terminology of the industry the skins of large animals, such as full grown cattle and horses, are called hides; the term skins is used when referring to small animals, such as sheep, goats, and calves. The term "hide" is never applied to the small animals.

Cattlehide

In the case of cattle, the protection of the animal is both from hair and from moderately heavy hide structure. The fiber of the skin is heavier in the back areas than in the belly, and the hair is longer. In the subclassification of cattle we can consider both beef cattle and dairy cattle. Beef cattle are raised primarily for their meat and, as such, are kept confined and fed a diet of high protein foods.

The development of fat beef cattle results in heavy fat deposits in the hides and a change in the fiber orientation. This "vertical fiber defect" is related to the problem of pulpy butts. Dairy cattle are kept under much less strenuous conditions than beef cattle, are adequately protected by barns, and are fed a diet which will aid in their milk production. As compared to beef cattle, dairy cattle have very thin, spready hides and shorter hair; their skins contain less fat, are not as thick, and have relatively large surface areas. The Holstein or Guernsey type dairy cattle are the most common in the United States.

Calfskin

Male dairy calves are slaughtered at the best economic age for reasonable yield in meat. The European practice generally is to kill the calves at a younger age than in the United States, As a result, European calfskins may be a bit smaller than domestic calfskins. Since the cattle hides described earlier are the result of maturing calfskins, we would expect the relationship between the two to be primarily one of size. It has been proved that the calfskins will have the same number of hair follicles as mature skins. The main difference, then, between calfskins and cattlehides, from a structural point of view, is the fineness of grain. Since the hair follicles are much smaller, collagen bundles have a very fine structure as compared to cattlehides and are useful for the finest of leathers.

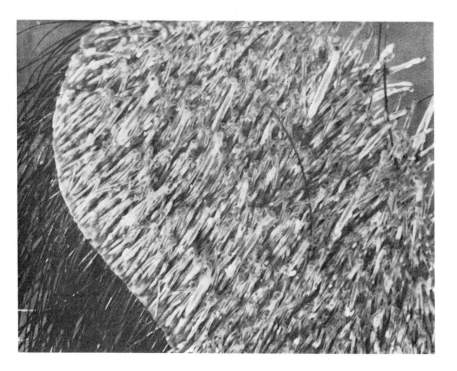

Figure 3 A cattlehide was prepared by enzymatic unhairing and the entire grain (or keratin) layer, including the hair, was removed intact. This remarkable photograph shows the hair roots and the structure of the keratin going down into the hair follicle. Adhering to some of the hairs are sebaceous or fat glands which lubricate it. (*Courtesy USDA*)

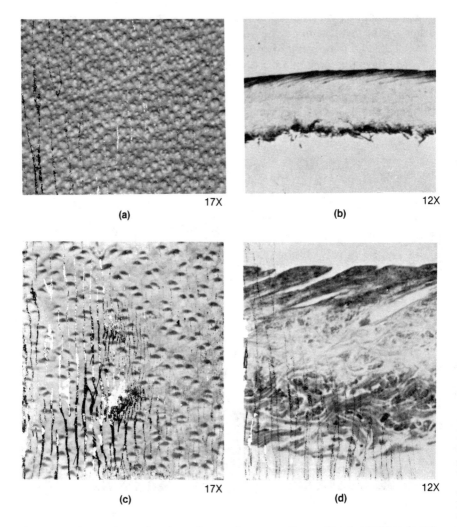

17X
(a)

12X
(b)

17X
(c)

12X
(d)

Figure 4 Comparison of grain surface and cross sections of bovine leather. In this comparison the calfskin clearly shows a much finer grain and closer hair follicle pattern to account for the fine grain of calfskin. Sole leather and side leather have the same grain pattern for hair, and they are also alike in grain pattern in the cross section showing same size grain area and hair follicles. The thinner upper leather is due to splitting. **(a)** and **(b)** Calfskin leather. **(c)** and **(d)** Side upper leather (steer). **(e)** and **(f)** Sole leather. (steer) (*Courtesy Tanners Council Laboratory*)

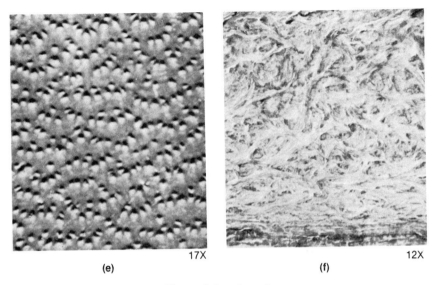

	17X		12X
	(e)		(f)

Figure 4: (continued)

Sheepskin

Sheep, quite unlike cattle, are raised primarily for their wool. Through selective breeding the hair growth has been improved, resulting in very long fibers and fine texture. Since the sheep is protected primarily by the wool, the function of the skin is more to support the growth of the wool than to serve as a protective organ in itself. In a cross section of sheepskin a large number of fat cells are visible. The oil given off by these cells will lubricate the wool. Sheepskin is very open and porous and has very little structural fiber. This lack, and the large concentration of glands in the area at the base of the hair root, result in a physical weakness of the skin at this point.

Goatskin

The goat is an ideal animal for tropic countries and localities where adequate grazing land for sheep or cattle is not available. The protection of the animal is partly from hair and partly from the fiber of the skin. Goatskins, as compared to sheepskins, have a very tight fiber structure and are easily recognized. The tight-natured fiber of goatskin allows its use in the more durable type of applications in the manufacture of gloves and shoes.

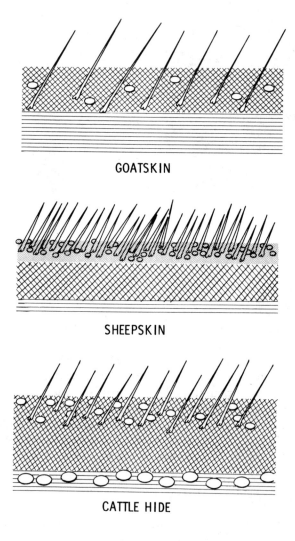

GOATSKIN

SHEEPSKIN

CATTLE HIDE

Figure 5 Structural comparison of hides and skins. These idealized drawings in
dicate the main structural differences between goatskins, and cattlehide. Goatskin
have less hair and less fat than sheepskin and have a tight, firm structure. Sheepski
contains many fat glands and hair roots and is an open type skin. Cattlehides hav
fats both near the hair root and near the flesh side; the structure is tighter than tha
of sheepskin but is more open than goatskin.

24

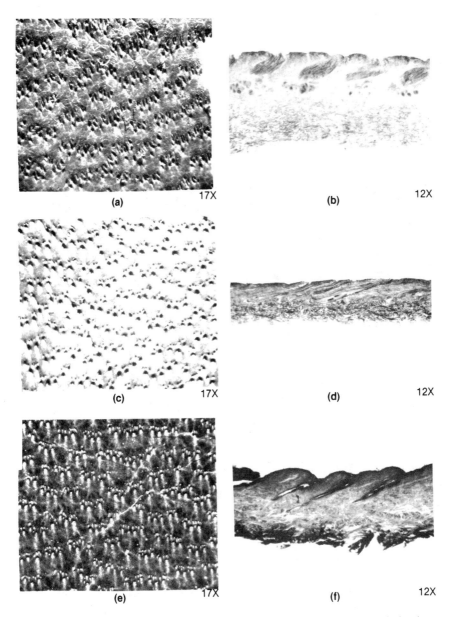

Figure 6 Comparison of grain and cross section of goatskin and sheepskin leather. The difference in the structure of wool sheepskin leather and hair sheepskin leather is in the size of the hair follicles and the density of the hair. The wool sheep is more porous; the hair sheep is much tighter; the goatskin has a wider pattern of hairs and a denser structure of skin. (**a** and **b**) Wool sheepskin leather. (**c** and **d**) Hair sheepskin leather. (**e** and **f**) Goatskin leather. (*Courtesy Tanners Council Laboratory*)

25

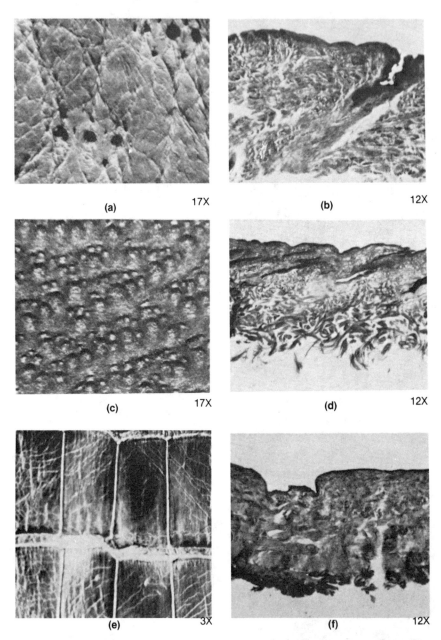

(a)   17X   (b)   12X

(c)   17X   (d)   12X

(e)   3X   (f)   12X

Figure 7 Comparison of grain and cross section of miscellaneous leathers. The pigskin cross section shows the hair follicle going completely through the skin and the grain showing the rough grain and grouping of the hair pattern of pig. Horse front has a structure resembling hair sheep or goat; it has a strong tight fiber. The alligator leather has the characteristics of reptiles: no hair follicles or fat glands, but a dense woven network of hide fibers. (**a** and **b**) Pigskin leather. (**c** and **d**) Horse front glove leather. (**e** and **f**) Alligator leather (7e is a 3-power magnification). (*Courtesy Tanners Council Laboratory*)

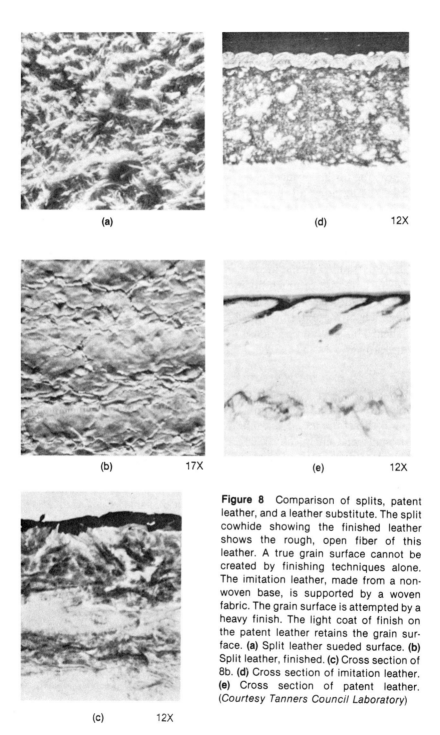

(a)

(d) 12X

(b) 17X

(e) 12X

(c) 12X

Figure 8 Comparison of splits, patent leather, and a leather substitute. The split cowhide showing the finished leather shows the rough, open fiber of this leather. A true grain surface cannot be created by finishing techniques alone. The imitation leather, made from a nonwoven base, is supported by a woven fabric. The grain surface is attempted by a heavy finish. The light coat of finish on the patent leather retains the grain surface. **(a)** Split leather sueded surface. **(b)** Split leather, finished. **(c)** Cross section of 8b. **(d)** Cross section of imitation leather. **(e)** Cross section of patent leather. (*Courtesy Tanners Council Laboratory*)

27

Pigskin

The structure of pigskin is in accordance with the habits of the animal. The domestic pig is protected by a layer of fat lying just below the surface of the skin. A pig has very little hair and its skin is of a relatively tough tight-natured weave with a large quantity of stored food fat. The hair of the pig is relatively stiff, is set in small clumps, and the bottom of the hair follicle is very near the inside surface of the skin. Pigskins, therefore, are essentially porous, having holes all the way through due to the hair follicles.

Horse

Horsehides may be divided into two parts. The forward section of the horse is relatively light skin and in spite of a fairly heavy growth of hair, the texture of this area is not much different from some types of goatskin. At the top of the hindquarters the skin is much thicker. Over the hindquarters and in the center of the skin fiber is a close network of fibers known as the horse butt shell. This shell, a very dense structure, is the source of the genuine cordovan leathers.

Reptile

Reptiles are cold-blooded animals. Their skins have no thermostatic function and therefore they are devoid of hair and fat glands. The scales are functionally and chemically related to the hair of warm-blooded animals. Histological investigation indicates the nature of the protection gained from the scales in this type of animal.

Fish have an entirely different skin structure from lizards and snakes in that the water protects the fish. Also, in the case of sharkskins the scales are very small and an outer, inert layer called shagreen is present to further protect the animal.

REFERENCES

Everett, A.L., Willard, H.J., and Naghski, J., *J. Am. Leather Chemists' Assoc.*, **61**, 112 (1966).

Everett, A.L., Willard, H.J., and Windus, W., *J. Am. Leather Chemists' Assoc.*, **62**, 25 (1967).

McLaughlin, G.D., and Theis, E.R., "The Chemistry of Leather Manufacture," ACS Monograph 101, New York, Reinhold, 1945.

Mellon, E.F., and Korn, A.H., *J. Am. Leather Chemists' Assoc.*, **51**, 469 (1956).

Mellon, E.F., Audsley, M.C., Viola, S.J., and Naghski, J., *J. Am. Leather Chemists' Assoc.*, **58**, 514 (1963).

Mitra, S.K., "Indian Hides and Skins, Histological Characteristics," Madras, Central Leather Research Institute, 1963.

O'Flaherty, F., Roddy, W.T., and Lollar, R.M., "The Chemistry and Technology of Leather," ACS Monograph 134, Vol. 1, p. 4, New York, Reinhold, 1956.

Roddy, W.T., *Boot and Shoe Recorder* (Sept. 1, 1966).

Wilson, J.A., "Modern Practice in Leather Manufacture," New York, Reinhold, 1941.

CURING

3

Hides are seldom processed directly into leather, but rather are held in a state of preservation and later tanned by specialists in the art. Whatever the source of the hide, or whatever the quality of the skin, it is on the world market. The value of the skin, local customs, and economic conditions will dictate the method of preservation to be used. In this chapter we will consider the theory, technology, and economics of the three major curing systems. These systems are drying, salt curing, and brine curing.

MAIN ASPECTS OF HIDE PRESERVATION

When the hide or skin is removed from an animal, the outside of the skin is covered with the dirt of the animal's surroundings. The inside of the hide or skin of a living animal contains bacteria and other microorganisms held in control by the metabolic defenses of the animal. When the animal is killed, the process of decay starts immediately.

We may consider the hide to be made up of (1) fibrous proteins, keratin, collagen, and elastin, (2) the soluble or soft proteins, albumins, globulins, mucoproteins, and soft keratins, and (3) the fat components of the body, both physiological and fat storing. Of these materials the most resistant to chemical and bacterial attack are the fibrous proteins, hair and elastin. The next most resistant is collagen. The least resistant are the mucoproteins, the albumins and the globulins. The soft body fats are also subject to bacterial decay.

Decomposition of a hide, therefore, will not proceed at the same rate in all parts. The initial decomposition will take place in the areas of high metabolic activity and in those containing the soluble materials. The blood in the veins and the soft proteins in the hair follicle will be the first sites of significant decomposition. Since the soft proteins in the hair follicle begin to decay first, the hair becomes loose and a "hair slip" hide is indicated. This would be classified as a Number 3 hide, of very poor quality. This initial decomposition and the evidence of it, is a good indication to the tanner of the state of the hide at the time of curing, and indicates whether or not bacterial decomposition has occurred. Rotten or stale hides can be smelled, but hides which have been given a delayed cure, i.e., decay has started prior to the initial salting, may not smell bad in the salted condition. Stale hides, or hides in the first stages of rotting, will have predominant blood vessels when made into leather. Decomposed areas of a hide may not be evident to the inexperienced inspector when the hides are in a cured condition. Such hides, when put into soak, will become even more damaged if the bacteria become active in the soak. The damage may be confined to grain pitting, but in extreme cases, large areas of the hide may be destroyed.

The second area of decomposition is the collagen itself. The initial stages of collagen damage will take place at the juncture between the grain layer at the root of the hair follicle and the corium. In this part of the hide there is a transition in the fiber structure and this is a site of some mechanical weakness. When the hair slip or decomposition has proceeded to this point, the entire grain down to the base of the hair root may peel back, leaving a rough surface.

Bacterial action on the fats will cause a breakdown of some of the unsaturated fats with a release of shorter chain fatty acids. This will result in a general yellowing of the fats, which will be softer and have a lower melting point

In addition to bacterial changes, others result from the nature of the protein system. Some of the soluble proteins will coagulate with heat or upon standing and will gel within the hide. Another change may be induced by metabolic enzymes within the hide itself, which continue their activity and bring about an autolytic decomposition of the hide under sterile conditions. Proper curing methods will minimize both bacterial decomposition and biochemical changes in the hide, so as to maintain good leather-making qualities.

Drying

Air drying of hides and skins is one of the oldest methods of skin preservation. It is the standard method of preservation of most reptile, goat, and fur skins and of a very large percentage of the hides and skins obtained in the tropical areas of Central and South America, Africa, and Asia.

The drying of fresh hides and skins is governed by the same physical laws applied to the drying of any dense, fibrous material. The rate of drying is determined by temperature and relative humidity of the air and the ability of the hide to give up moisture. The general principles governing the drying of hides are the same as those which apply to the drying of leather, as covered in this book.

Air drying has the advantage of being the simplest form of hide and skin preservation. It is most practical in areas where salt is expensive and a small number of skins are to be cured. It is particularly useful in dry, tropical climates and in the rural areas of some underdeveloped nations.

The rate of drying is of great importance. If it is too slow, putrefaction may begin in the hide before the moisture level can be decreased to the point

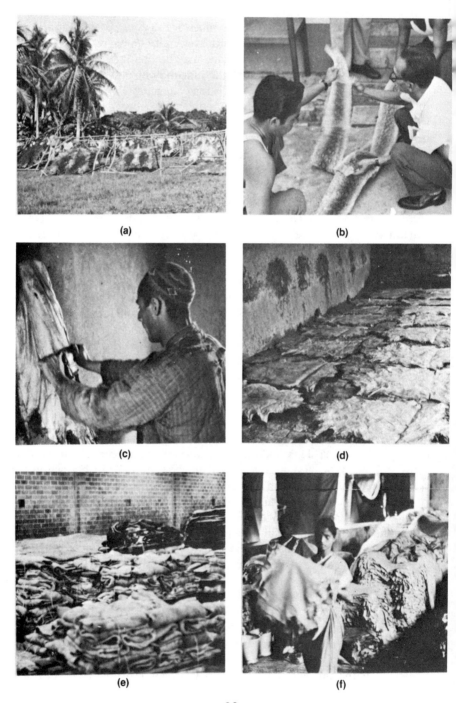

(a)

(b)

(c)

(d)

(e)

(f)

32

where bacteria can no longer be active. If drying is too fast, the outer surfaces of the hide may become hard and dry while the inner parts still have enough moisture to support bacterial growth. This "case hardening" of the outer surface prevents the proper removal of moisture, and the hides may rot out from inside. Upon soaking for processing, the hide will have a blistered appearance and eventually holes will appear.

The technique of drying hides and skins varies, depending upon the circumstances of the area, the weather conditions, and the availability of auxiliary materials. The simplest method of drying hides is to spread them on the ground, supported on sticks or stones. Even though this method of supporting the hides does not allow the skin to lie flat, it is necessary to prevent exposure to the ground and being severely damaged by insects. More often the hides are hung on poles or ropes and dried in the sun. This gives better ventilation and more uniform drying. However, the skin may be damaged from the permanent fold marks of hanging, and direct sunlight may cause case hardening of the upper surface. Frame drying is by far the most common practice. The hide is stretched on a frame and allowed to dry either in the shade or in the sun. As the hide dries, it shrinks and tightens out flat. This allows easier packing and good circulation of the air around the hide during drying. In a hot climate it is better to dry the hides in the shade, preferably in an open-sided shed. This also protects the hides from damage from rainfall.

Skins that are dried without any additional salt or other curing agent are classified as flint-dried skins. In those areas where salt is available, the skins may be dipped in a brine solution prior to drying. The bacterial action of the skin is held to a minimum by the salt during drying, and the salt also prevents coalescence of the fibers. The local salts used for this purpose may have distinct chemical effects on the hides, some of which may not be desirable.

As an aid to drying and prevention of insect damage, the skins may be soaked in a solution of arsenic salts, either prior to drying or near the end of the drying process. If the skins are to be folded before baling, they are usually folded just before the skin has reached its final dry state. The folding of very hard skins will cause grain crack and breaking of the fibers.

Figure 1   **(a)** Air drying of cattle hides in Malaya. **(b)** Inspecting air dried snakeskins in Singapore. **(c)** Hand fleshing of sheepskins. **(d)** Drying of salted sheepskins in Afghanistan. **(e)** Conventional salt curing of cattle hides. **(f)** Wet blue goatskins for export in India. *(Author's photographs.)*

Salt curing

The most common method of preserving cattle hides in Europe, North America, and other temperate climates is the conventional salt cure system. This method is employed by packing plants and hide dealers on a large scale. The fresh hide, direct from the kill floor, is taken into the hide cellar where it is put into a salt pack. The hides are laid flat in the pile and folded at the edges so as to keep the moisture in the salt pack as much as possible. Salt is strewn over the hide so that there is salt between each pair of hides. Approximately one pound of salt is necessary per pound of hide. The hide cellar is preferably a cool area where the seepage of brine coming from the hides during the curing can drain off. A salt pack will be built approximately twenty or thirty feet on a side and to a height of four or five feet. It may contain approximately two thousand hides, depending on the size of the salt pack and the average weight of the hides. Proper salt curing, in accordance with the Commodity Exchange rules, requires twenty-one days in the pack. If the pack is being made by a very small packer or by a collector of country hides, it may take several weeks to build the pack completely; hence the time of curing may not be the same from the top to the bottom. The twenty-one day curing time should be gauged from the day the last hide goes into the pack.

At the end of the curing time, when the hides are to be sold, they are "taken up" by a temporary crew. They are taken out of the pack individually and the excess salt is shaken from them. They are then thrown on the floor for inspection by representatives of the buyer and the seller. The hide buyers have a broom and a knife in hand to sweep off the excess salt and trim off misshapen hide pieces or other undesirable adhering flesh not normally considered part of a hide shipment. Inspection for Number 1's, 2's, 3's, and cuts and grubs is made at this time. The hides are then bundled and weighed for shipment.

The normal commercial practice of inspection of hides at the packing plant has been greatly modified through the years. Packing plants now seldom kill a wide variety of animals. The packer buys cattle to meet his orders for beef. Since the packer has standardized on the type of cattle processed to meet his customer demands, the hides from a particular source are generally predictable in type. Hide curing systems have been standardized so the buyer and broker have a good knowledge of what to expect from a particular source. Inspection at the source is, therefore, not often cost-effective. If a shipment of hides is, in the opinion of a purchaser, not of the quality purchased, a claim may be made against the seller. Since both the buyer and seller wish to continue to do business with one another, such claims are usually settled with a minimum of difficulty.

Hides that have been properly cured in the conventional manner will keep without refrigeration at moderate temperature in cool cellars for as much as a year. Since ideal storage conditions are seldom available, and maintaining an inventory of salted hides is seldom profitable, most salted hides are processed by the tanners soon after receiving them.

The techniques of properly green-salting hides require the use of good, clean salt. However, to purchase new rock salt for every new pack of hides would be expensive, so it is common practice to re-use the salt. This is done more extensively in the smaller packing plants or in the collector hide area where the quality of the hides is relatively poor and the economics of the business would not warrant the use of new salt. Since the negotiated price of any shipment of hides is based on the quality reputation of the source from previous shipments, the extensive re-use of old salt may be a false economy in anything but the smallest collector hide installation.

Brine curing

There are several systems of brine curing. The most significant at present is that used for the heavy packer hides in the United States. The advantage of brine curing to the packer is, primarily, that the hides can be processed in less time; therefore, less capital is tied up in hide inventory. The capital necessary for the equipment for a brine curing installation can usually be obtained from the liquidation of the hide inventory in the salt packs. The hides received directly from the killing floor are sent through a chilling machine to clean off some of the manure and surface dirt. They are then taken to a whole hide fleshing machine equipped with a demanuring and fleshing cylinder. Since the body temperature of the hide has been decreased to a point where the fats are hard, the fleshing operation can be effectively accomplished. The manure is washed away and removed more thoroughly with the demanuring cylinder, and the flesh can be recovered and rendered for fat, soap, or fatliquoring oils. The fleshed hides are then soaked in brine, either in vats or in a raceway system. The hides are cured for forty-eight hours or more in a saturated brine medium; they are considered cured when the saturated brine has completely permeated them. The hides are removed and wrung on a wringing machine, inspected, and a small amount of additional safety salt is added. They are then bundled and shipped.

The fleshing and trimming operations result in an approximately 22% decrease in the weight of the hide from what it would be in conventional curing. This, of course, is reflected in the price of the hides. Fleshing of the hides increases the rate at which the curing operation can be conducted, and also yields valuable fat to the packer. Prefleshing of the hides prior to cure

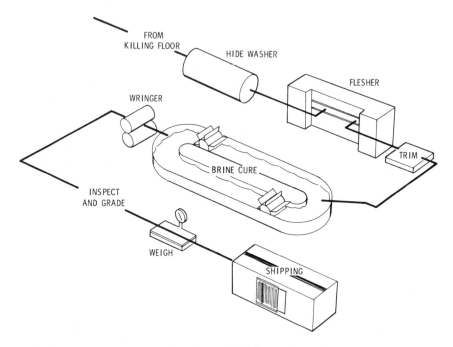

Figure 2 Process flow drawing of a typical brine curing system.

Figure 3 Hide washing machine. The picture shows an installation of two hide washing machines, showing the pegs inside the machine for the tumbling of the hides, the motors for the drive mechanism, and the pipe into the washing machine for spraying cold water. At the discharge end of the machine the hides will come out onto a conveyor. In this case the conveyor belt is made up of sections of wood and the hides will be carried wet and chilled to the fleshing machines. *(Courtesy Chas. H. Stehling Co.)*

(a) Back.

(b) Front.

Figure 4 Whole hide fleshing and demanuring machine. Two views are given for the whole hide fleshing machine. This machine is one of the most significant advances in the hide cure systems used in the United States. The machine is designed to remove the flesh and the manure simultaneously. The fresh hide, chilled from the washing machines and discharged at this machine, is handled by two operators. Half of the hide, either front or back, is fed in first; then the hide is turned and the remaining half is put in. The demanuring cyclinder is down low in the front of the machine and can be seen in the front view at the bottom of the machine. This cylinder moves into the machine with the motion of closing jaws. The manure being knocked off at this point is kept separate from the flesh, which is taken off by the sharp fleshing cylinder. The flesh can then be recovered separately from the manure and sent to the rendering plant. At the back of the machine there is a track which will hold a grinding stone that can be moved back and forth across the fleshing cylinder to sharpen the blade. *(Courtesy Chas. H. Stehling Co.)*

is an advantage to the tanner in that the soak fleshing operation may be eliminated and that there is less freight cost. In an efficient fleshing operation, and with normal prices for the fats, the cost of the fleshing operation would be paid for by the grease obtained.

The success of brine curing in the United States has resulted in the installation of a large number of brine plants by packers and by hide dealers who may service several packing plants. Brine curing, however, is not as effective as conventional salt pack curing with respect to the effectiveness of the preservation of the hides. Brine cured hides, therefore, are not as safe for long periods of storage as conventionally cured hides. Under normal circumstances, their state of preservation is sufficient for six months or more, and usually the hides are satisfactory when shipped for export.

Short-term preservation

In the United States there has been an increasing trend towards the location of packing plants near the source of the animal slaughter to eliminate the freight and also to obtain production plants in areas of available labor supply. Modern U.S. cattle packing plants generally kill approximately 1,000 head per day. These units operate in the meat producing areas in the midwest and southwest of the United States. Many of these plants are equipped with brine curing units while others are processing hides by conventional cure.

A third method of handling the hides from these and smaller units is to ship the hides to a hide processing plant for curing or leather production. In shipping and delaying cure or processing, there is a distinct possibility of the development of bacterial damage.

Fresh hides, if immediately removed from the animal, can be chilled and washed to remove any excess manure and then sprayed with a disinfectant. The longer the period of time the disinfectant is effective the longer it will be before there is any damage to the hides and the greater the distance that the hides can be shipped. There are distinct advantages in the United States to have preservation for 24 hours which would allow shipment up to 500 miles without damage to the hides. If preservation could be relied upon for 14 days under warm conditions, without the preservation hides could be shipped from any point in the United States to any other point for commercial processing with danger of bacterial damage. Under these circumstances, there would be no need for the salt cure and the additional problem of salt pollution would be eliminated. Short-term preservation should eventually replace the conventional salt cure of brine cure operations as a major means of commercial handling.

A small packing plant of 1,000 hides per day is marginally small for the efficient operation of a tannery. A tannery operating larger levels can gain greater efficiency but requires drawing hides from a larger radius.

Figure 5 Raceway brine cure tank. This oval shaped tank is kept agitated by the use of two small paddles. The salt concentration is kept near saturation, automatically. *(Courtesy Diamond Crystal Salt Co.)*

Figure 6 Hide wringing machine. This machine is a feedthrough unit used in brine curing systems, and as indicated, there are two belts which feed the hide in through the rollers. Pressure on the rollers is controlled hydraulically so that constant pressure can be applied regardless of the thickness of the hides being processed. *(Courtesy Chas. H. Stehling)*

The use of disinfectants adds to the environmental problems of both the packer and the tanner. A practical solution is to chill the hides with cold water, drain them, and ship the hides under refrigeration. This system is effective for only a few days depending on the climate and the temperature of the refrigeration. Cooperation of the packer with the tanner is essential.

WET BLUE OR BLUE SIDE

The practice of processing hides through the chrome tan in tanneries near the sources of the hides is the most significant development in the commercial aspects of the leather industry in the past quarter century. The hides are beamed and chrome tanned in very large tanneries. The chrome tanned hides may be shipped to an affiliated tannery in another area, or may be sold. Some large "blueing" tanneries also work as contract tanners and are paid for processing hides for others. Wet blue production of cattlehides is in rapid growth in the United States at the present time. We can expect growth in wet blue operations in the United States in the next decade. It is expected that the next probable step will be to a neutral crust, which is particularly desirable for international shipment because of the lower weight per piece. There are considerable advantages to processing hides at the source. These advantages include saving on the cost of curing, freight, potential quality selection, environmental considerations, and better preservation of the hides. Either the brine cure system or the salt pack cure system results in a preservation cost of approximately four dollars per hide.

Hides may be unhaired by the hair burn system in a hide processor, followed by bate pickle and tan. This can be done without removal of the hide from the processer at any point in the operation. Upon completion of the unhairing pickle and tan, the hides are wrung and sided. Sides may be inspected and classified at the time of the wringing. Blue sides may be palleted and sent to the tannery for further processing.

Some blue side companies operate on a contract basis and may process hides in through the blue for other tanners or for dealers for export.

ECOLOGICAL CONSIDERATIONS

With the development of ecological responsibilities as the result of changing times, the problem of processing, unhairing hides, and getting rid of the large quantities of liquid and solid wastes has become acute in urban areas. In the Midwest and West of the United States, hides are produced in the agricultural areas. These areas have more land available for sludge disposal, and the objection by neighbors to the presence of a tannery is almost eliminated.

LOOKING AHEAD

Trends are well established throughout the world for more processing of hides at the source. This approach results in a more efficient utilization of by-products and a decrease in the weight to be shipped without the possible loss of valuable leather substances. These trends will probably continue and we can expect the development of blue side and crust operations throughout the world to introduce new categories of commercial types of leather.

REFERENCES

Tancous, J. J., Roddy, W. T., and O'Flaherty, F., "Skin Hide and Leather Defects," Cincinnati, Western Hills Publishing Co., 1959.

U.S. Department of Agriculture, "Technical-Economic Evaluation of Four Hide Curing Methods," 1962.

Wilson, J. A., "The Chemistry of Leather Manufacture," Vol. 1, New York Reinhold, 1928.

Mann, I., "Processing and Utilization of Animal By Products," Food and Agriculture Organization of the United Nations, 1962.

HIDE AND SKIN PROTEINS

4

The production of leather can be looked upon as taking place in three steps. The first step is the removal of the unwanted components, hair, fats, etc., leaving a network of fibers of hide protein. The second step is to react this network with tanning materials to produce a stabilized fiber structure. The third step is to build onto the tanned fibers characteristics of fullness, color, softness, and lubrication, and to finish the fiber surface, to produce a useful product. The chemistry of the hide protein determines the reaction of the tanning process as well as the chemistry of the tanning materials. An equally important factor is the physical condition of the fiber, which is due to the swelling properties of the protein. An elementary understanding of protein chemistry is necessary to appreciate and understand tanning reactions.

The subject of the biochemistry of hide protein is difficult when approached without a sound background in chemistry. For those readers who find this chapter difficult to understand in detail, it is suggested that it be read over quickly to serve as general background for the more descriptive approaches used in other parts of the book.

PROTEINS AND AMINO ACIDS

All living matter contains proteins. They occur in nature in a number of different physical forms and with a wide variety of chemical characteristics. All proteins are composed of long chains of amino acids, which are the essential components of living cells:

Amino acid   *Amino Acid Chain*

All amino acids, and consequently all proteins, are made up of the elements carbon, oxygen, nitrogen, and hydrogen. Sulfur is also found in some proteins.

In the complex life processes the sequence of amino acids in a protein is not random but, rather, is in a specific order. Although this sequential order contributes to some extent to the chemical behavior of the protein, the chemical characteristics of a protein are dominated by the nature of the amino acids.

42

The simple amino acid glycine can be used as an example of the general chemical behavior of amino acids and proteins. The compound contains both an acid and a basic group, and as such can undergo both acid and basic reactions.

Acetic acid

(a)

$$CH_3-COOH \rightleftharpoons CH_3-COO^- + H^+$$

at pH 3 to pH 5

Methyl amine

(b)

$$H_2O + CH_3-NH_2 \rightleftharpoons CH_3-NH_3^+ + OH^-$$

at pH 9 to pH 11

Glycine

(c) in acids

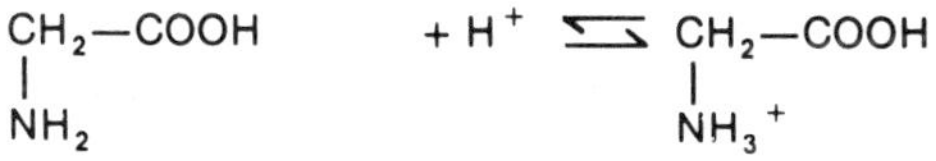

$$\underset{\underset{NH_2}{|}}{CH_2-COOH} + H^+ \rightleftharpoons \underset{\underset{NH_3^+}{|}}{CH_2-COOH}$$

+ charge   acid binding

(d) in bases

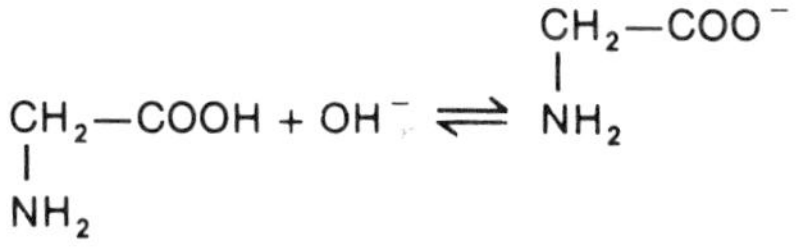

$$\underset{\underset{NH_2}{|}}{CH_2-COOH} + OH^- \rightleftharpoons \underset{\underset{NH_2}{|}}{CH_2-COO^-}$$

− charge   base binding

(e) at isoelectric point

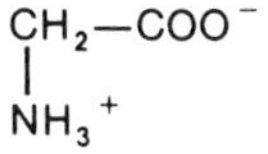

$$\underset{\underset{NH_3^+}{|}}{CH_2-COO^-}$$

zwitterion, zero net charge

The first reaction (a), the acid reaction, is a typical weak acid ionization, and this acid can be titrated in the acid pH range between about 3 to 5. The second reaction (b), the basic reaction, is typical of an amine and can be titrated in the alkaline range between about pH 8 to 11. In the third reaction, glycine, the amino acid has both the characteristics of acetic acid and methylamine. Under acid conditions, the ionization of the acid group COOH is minimal, but the NH₂ group absorbs the hydrogen ions, thus resulting in a positive charge and acid binding characteristics. At higher pH in the alkaline range, the presence of OH ions in the solution draws the hydrogen ion from the carboxyl group and also removes the hydrogen ion from the NH₃ + group. The molecule becomes positively charged in the acid range (c), zero charged near neutral (c), and negatively charged under alkaline conditions (d). As a result, the glycine has both a positive/negative charge on the same molecule. At some pH point, these charges will exactly equal one another. This point is the isoelectric point.

If the glycine is in a protein chain, both the acid and amino groups and carboxyl groups are tied up and cannot contribute as strongly to the acid base reactions of the protein. The acid base properties of the protein must then be due to the free acid groups and basic groups of the R component of the amino acids.

Proteins behave similarly to the glycine in that the protein is made up of a long chain of amino acids. With both acid groups and basic groups the protein will absorb hydrogen ions or give up hydrogen ions, depending upon the pH of the solution. As a result, the charge on the protein will vary from positive to negative and the protein will absorb acids in the low pH range and absorb alkalies in the high pH range.

The nature of amino acids varies quite widely depending upon the kind of chemical groups attached to the alpha-carbon atom. To date approximately eighty naturally occurring amino acids have been found. However, only about twenty of these are of major significance; these occur in protein materials to such an extent as to be considered major chemical components. These amino acids can be grouped into general classifications:

Amino Acids

Nonpolar

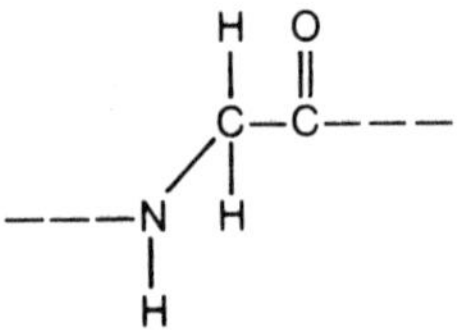

Glycine 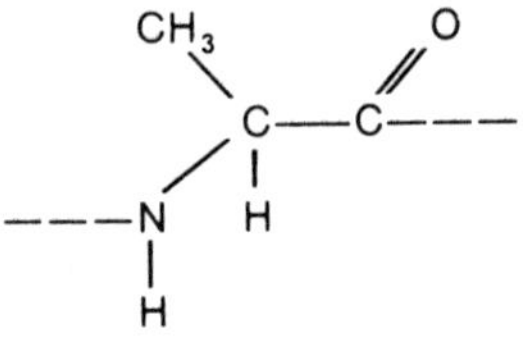   Alanine

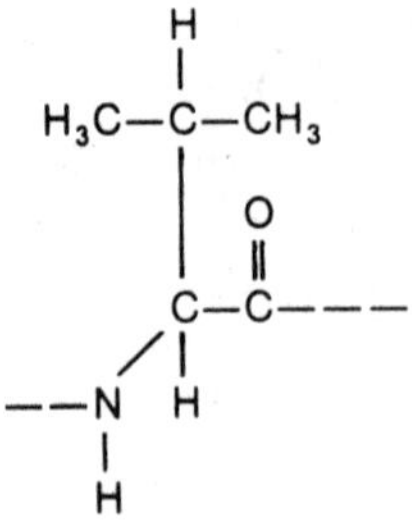

Valine 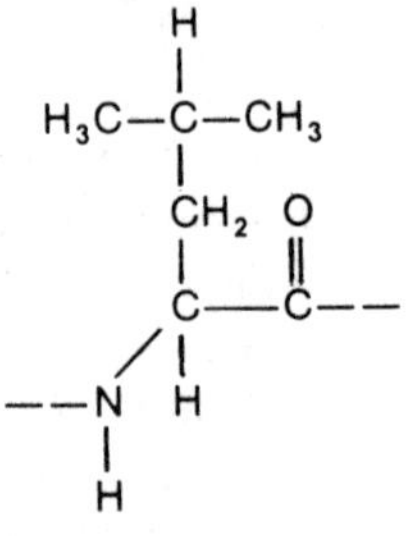   Leucine

STRUCTURE OF PROTEINS

Amino acids have the general structure shown. They differ from one another depending upon the nature of the side chain located on the alpha-carbon atom. If these side chains contain no polar groups, the acids are called nonpolar. If they contain a free carboxyl group or a hydroxyl group, they are classified as acid amino acids. If they contain amino groups or other nitrogen containing groups, they act as bases and are classified as basic amino acids.

Amino Acids

Acid

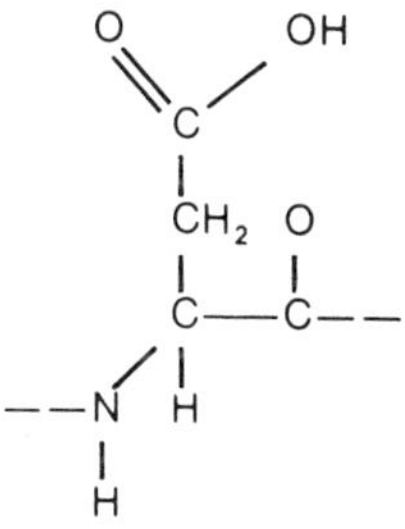

Aspartic acid 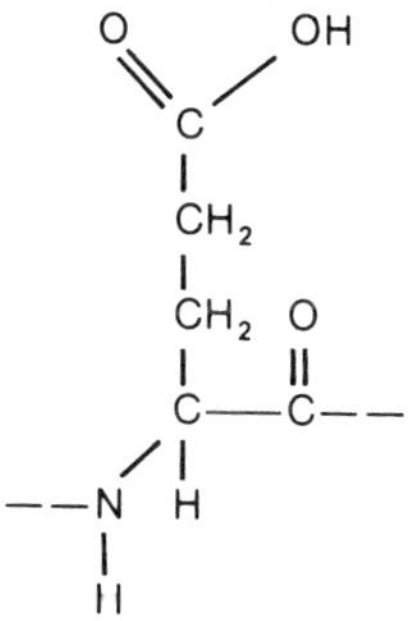   *Glutamic acid*

Amino Acids

Basic

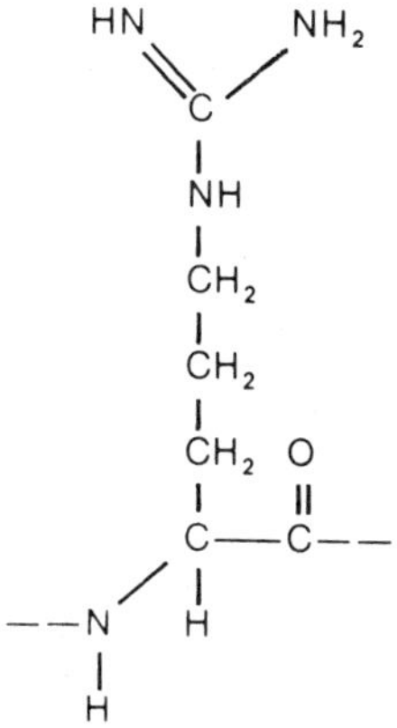

Arginine 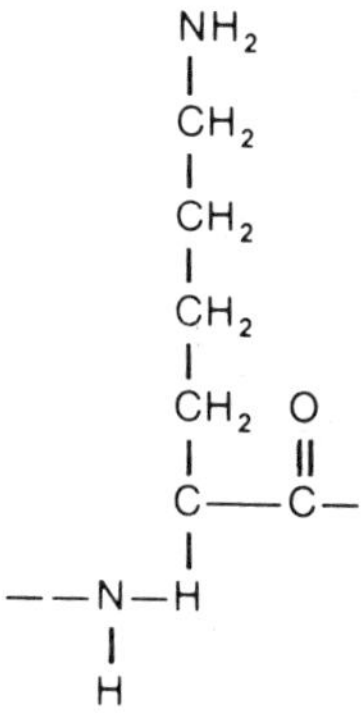   *Lysine*

The neutral amino acids, glycine, etc., are amino acids which have no free carboxyl groups or amino groups, i.e., the component fastened to the alpha-carbon atom is a hydrocarbon-type compound that has no acid nor basic reactive groups. Acid amino acids, primarily of aspartic acid and glutamic acid, are amino acids which have free carboxyl groups in their side chains and which will contribute to the acid character of the protein in which they occur. Basic amino acids are those in which there is available in the side chain a nitrogen component which is capable of acting as a base. The amino acids arginine and lysine are the most common of the basic amino acids. Such amino acids contribute to the basic character of the protein.

Amino Acids
  Others

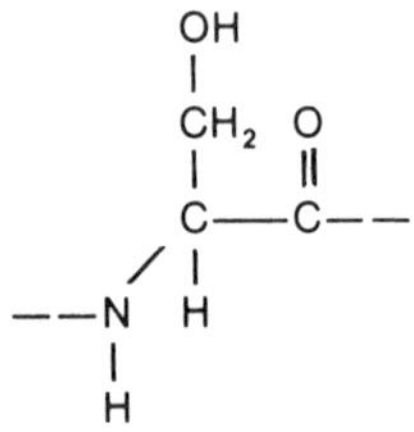

Serine

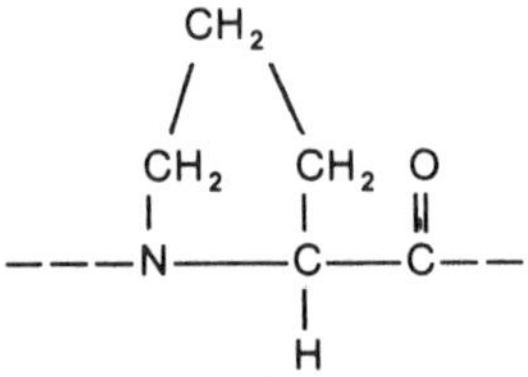

Proline 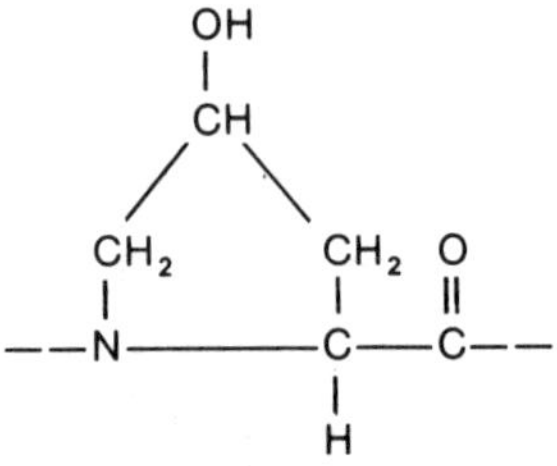   Hydroxyproline

In addition to the above three classifications there are other amino acids, each with specific chemical characteristics, which must be classified separately. Some of these may contribute to the specific chemical nature of certain types of proteins.

In their acid-base relationships the proteins behave like amino acids. Since the protein will contain both acid and basic amino acids, it will have ionizable acid and basic groups. These groups become ionized and bind acids and bases in accordance with pH conditions. The charge on the molecule changes from plus in the acid range to neutral due to the mutual balancing of the acid and alkaline positive and negative charges; to a negative charge in the alkaline range. In the case of a protein, a large number of different acid and alkaline groups may be present, each operating independently of one another. At some point on the pH scale the number of acid groups and alkaline groups which are ionized will balance one another and the protein is said to be at its isoelectric point. It is interesting to note that much of the chemical reactivity of hide during processing can be attributed to the acid and basic characteristics of the protein involved, and to the changes in the charges on the protein.

CLASSIFICATION OF PROTEINS

The amino acid composition determines the chemical and physical characteristics of a protein. Table 1 shows the amino acid content of proteins and Figure 1 shows an idealized representation of proteins.

TABLE 1

Amino Acid Content of Proteins (per cent).

Protein	Collagen	Elastin	Keratin	Albumin
Nonpolar				
Glycine	20	22	5	2
Alanine	8	15	3	6
Valine	3	12	5	6
Leucine	5	10	7	12
Other	4	15	7	9
Total	40	74	27	35
Acid				
Aspartic acid	6	.5	7	11
Glutamic acid	10	2.5	15	17
Total	16	3	22	28
Basic				
Arginine	8	1	10	6
Lysine	4	.5	3	13
Other	2	.5	1	4
Total	14	2	14	23
Others				
Serine	2	1	8	4
Cystine	—	—	14	—
Proline-Hydroxyproline	25	15	6	5
Other	3	5	9	5
Total	30	21	37	14

Albumin

Amino acid composition of albumin is given in Table 1. Albumin is one of the soluble proteins characterized by having a high percentage of acid and basic amino acids. The protein is highly ionized over a large portion of the pH range, and there are a large number of charges per unit weight. These charged sections of the protein have some electrostatic attraction for one another giving the molecule a tendency to fold back upon itself. This forms what are called molecular globules; such proteins are known as globular proteins. If a solution of a globular protein, say albumin, has added to it high percentages of salt (over 10%), the salt may act as an electrochemical bridge between adjacent protein molecules and result in the precipitation of some of the globular proteins. On the other hand, a lower concentration of

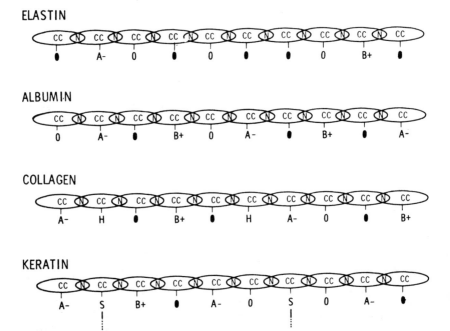

Figure 1 Idealized representation of the amino acids of the proteins:
Elastin has few acid amino acids (represented by A) and few basic amino acids (represented by B). There are many neutral amino acids (represented by ●), and the balance are in the "other" classifications represented by 0.
Albumin has many acid (A −) and basic (B +) amino acids. This chemical reactivity makes this a water soluble, non-fibrous, "globular" protein.
Collagen (leather-making protein) has more acid (A +) and basic (B +) amino acids than elastin but less than albumin. The hydroxyproline (H) is characteristic of collagen.
Keratin (hair type protein) is stabilized by cystine (− S −). This stabilizes its fibrous structure.

salt, in the neighborhood of 5-10% or lower, depending upon the protein, may aid in the solubilization of the proteins. For this reason, in the process salt curing, some of the globular proteins are leached out by the salt.

If a solution of albumin is raised in temperature to its critical point, the molecular agitation may become sufficient to overcome the intramolecular electrostatic attraction and allow the adjacent molecules to come into contact. At this point cross linking between molecules may occur and essentially the whole protein solution may become one large molecule through mutual intermolecular electrostatic attraction. This is the reaction that is observed in the insolubilization of albumin in the boiling of an egg.

Elastin

Elastin is one of the proteins that has very few acid and basic amino acid groups. Elastin differs from albumin in this regard and is remarkably inert to chemical attack. If we place a piece of freshly cured hide in a solution of 0.1 N hydrochloric acid and boil it under reflux, the hide structure will be destroyed, but a network of elastin will be left intact. Even under the conditions of strong acid, the elastin network does not have sufficient charge per unit mass to be solubilized. Silk has an amino acid content which is somewhat similar to elastin, and as a result they share many of the same characteristics. The fibers do not have a tendency to stick together, even upon drying, and they do not need chemical stabilization or tanning to prevent their decomposition.

Collagen

The protein collagen contains both acid and basic amino acid groups, as well as nonpolar amino acids and a relatively high percentage of proline and hydroxyproline. For present considerations, we can assume proline and hydroxyproline to be inert. From the amino acid content of collagen we see that, with regard to chemical reactivity, this material lies between elastin and the globular proteins. In neutral solutions collagen is insoluble, and unless subject to heat or bacterial degradation, collagen will not break down in water solution. When strong acid is added collagen may be dissolved; it is not as inert as elastin in this regard. Collagen may also be dissolved by strong alkalies.

It would be possible, therefore, by properly selecting conditions of salt and acid content, to separate the albumin, elastin, and collagen components of a hide. These factors are of major importance in the manufacture of leather.

The number of amino acids comprising a protein will depend upon the nature of the protein. Measurements have indicated molecular weights of

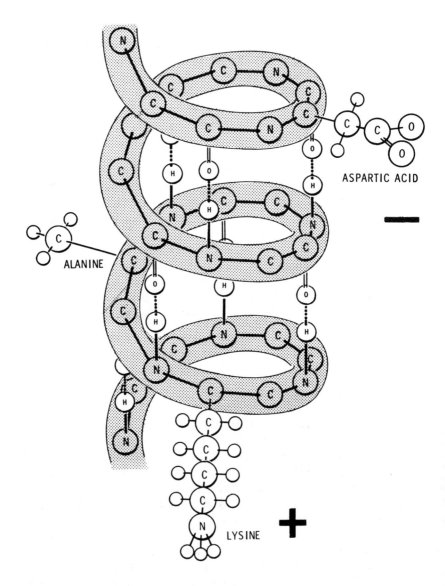

Figure 2 Pauling proposed the alpha helix structure for proteins. This drawing is an idealized representation of such a structure. The coil is held in place by the hydrogen bonding. The "R" groups of the amino acids are arranged as appendages on the main coil. The aspartic acid and lysine are shown in the normal ionized state.

ASPARTIC ACID

ALANINE

LYSINE

protein range from about 10,000 to 200,000 or more. In the case of hide collagen, the accepted lowest molecular weight is of the order of 40,000. This would indicate that the number of amino acids in the single molecule would be approximately 500–1,000. The amino acids are not in a random order, but are in a definite pattern. In the ability of the living cell to place the amino acids in a specific sequence, most probably lies the secret of life and heredity.

Keratin

The amino acid composition of keratin shows more acid amino acids than collagen, but less than albumin. We would expect, therefore, that keratin (hair protein) would be less soluble than albumin, but slightly less stable than collagen. This, of course, we know is not true. Collagen may be solubilized by hot water, acids, and alkalines. Keratin is very stable to hot water and mild chemical attack.

The stability of keratin is due to presence of cystine. Cystine is an amino acid containing sulfur. This sulfur forms a stable sulfur-sulfur linkage between protein hairs.

PHYSICAL STRUCTURE OF PROTEINS

In addition to the definite order of the amino acids, there is the question of the physical orientation of the amino acids to one another. It is possible by means of X-ray studies to determine the spacing of the atoms of a molecule. Extensive studies have been made of the structures of proteins by electron microscopy and by X-ray scattering data. Through these studies the dimensions of the amino acids in the polypeptide and protein chains have been determined with considerable accuracy. From this it has been calculated that a molecule of a protein consists of a spiral chain of amino acids. The calculations of the interatomic distances have been used to devise working models of the protein structure. The individual protein then appears to be a coiled spring with the R groups of the protein sticking to the outside, like appendages. We have, therefore, a barbed wire type of structure, and the carboxyl and amino groups of the amino acids are oriented to the outside. The availability of these carboxyl groups, amino groups, and other reactive groups to form hydrogen bonds and to be attracted to one another leads to a further definite orientation of the protein molecules. There are individual strands of amino acids which are coiled together to form a larger coil, which is the protein chain. These larger segments, in coils of three, then pack together and form the protein fibrils. An excellent summary of the relationship of amino acids and proteins into the structure described above is given by Highberger in his John Arthur Wilson lecture to the ALCA in 1960.

In spite of the complicated fiber structure, the orientation of the protein coils, and the location of the amino acids, the distances between the amino acid chains are such that the reactive materials (e.g., tanning materials and water) can pass relatively freely through the spaces between the protein chains. Thus, protein can react chemically in essentially complete stoichiometric relationships with small molecules.

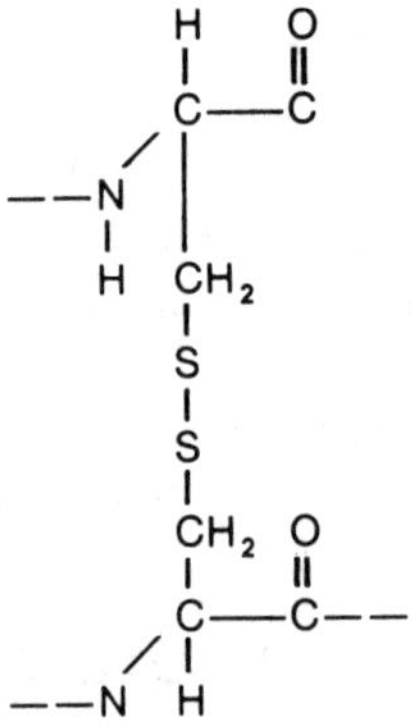

$$
\begin{array}{c}
\text{H} \quad \text{O} \\
\mid \quad \parallel \\
\text{C}\!-\!\text{C} \\
\diagup \quad \mid \\
-\!-\text{N} \quad \mid \\
\mid \quad \mid \\
\text{H} \quad \text{CH}_2 \\
\mid \\
\text{S} \\
\mid \\
\text{S} \\
\mid \\
\text{CH}_2 \quad \text{O} \\
\mid \quad \parallel \\
\text{C}\!-\!\text{C}-\!- \\
\diagup \;\mid \\
-\!-\text{N} \quad \text{H}
\end{array}
$$

Cystine
(cross linking two chains)

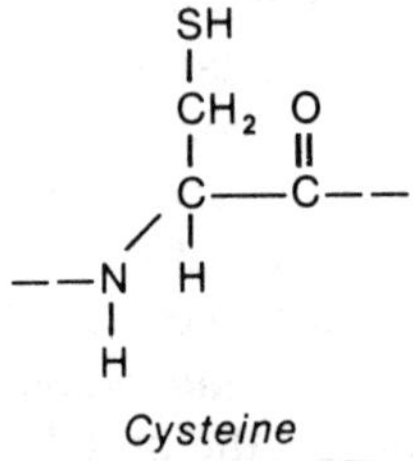

$$
\begin{array}{c}
\text{SH} \\
\mid \\
\text{CH}_2 \quad \text{O} \\
\mid \quad \parallel \\
\text{C}\!-\!\text{C}-\!- \\
\diagup \;\mid \\
-\!-\text{N} \quad \text{H} \\
\mid \\
\text{H}
\end{array}
$$

Cysteine
(reduced *cystine*)

The individual protein chains are not entirely separate chemical entities, but there may be a cross linking of the amino acid chains to one another by certain bridging types of amino acids. The most important of these, from the leather technologist's point of view, is cystine. This amino acid has two cysteine amino acids in adjacent polypeptide chains cross-linked by a sulfur bond to form cystine in a stable fiber structure.

It is generally accepted that the fibrous stability of the protein structure is due to the hydrogen bonding and electrostatic affinity from one chain to another. When a fibrous protein is subject to extensive chemical action and hydrolysis, the hydrogen bonding between polypeptide links may be broken and additional water molecules absorbed. Thus, the treatment of protein with hydrolyzing chemicals will first disrupt the linkages between the chains bringing about the protein dispersion. Once the protein is dispersed, more acid and basic groups become available and, as a result, the acid-base binding capacity of the protein is increased. This may occur even with very little change in the physical appearance of the fiber.

To break the protein down to its individual amino acids requires extreme chemical action, such as prolonged boiling with acids or alkalies. Specific enzymes may also bring about a cleavage between individual amino acid links, causing a change in the physical structure of the protein. The strength of the bonding of the proteins is very high; properly oriented protein fiber is many times stronger than steel. We cannot, of course, obtain such high theoretical strength in leather or a pure protein fiber because the ultimate strength of the leather is due to the strength of the fibers and their ability to hold together. It would be rather unlikely that the physical breaking of a piece of leather would be due to disrupting of the polypeptide links of the amino acids. The individual protein fibers and fibrils are formed together into an organized pattern of fibers which are intertwined in an apparently random fashion.

SWELLING

Just as important as the acid-base binding capacity of the protein is swelling. No other phenomenon is of such great importance to practical leather manufacture. Swelling affects the size and shape of the hide fiber, the nature of the grain, and leather quality. It is, therefore, of great importance that the mechanism of swelling and the means of controlling it during leather processing are understood.

Consider the physical condition of a hide in water. For this purpose, assume the hide to be a uniform solution of gelatin on one side of a membrane standing in a solution. We could also consider the protein to be a brush

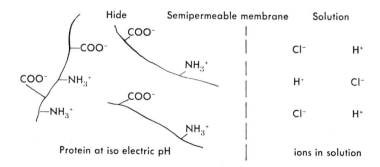

heap type structure of protein molecules with no membrane involved, except in a mathematical sense. By either approach, the same results are obtained.

The addition of hydrochloric acid to the gelatin will result in fixation of the hydrogen ions by the gelatin. Consequently, there will be a migration of the hydrogen chloride from the solution into the gelatin phase through the real or imaginary membrane. Since most of the hydrogen ions are fixed by the protein, more hydrogen ions will be attracted into the gelatin phase, and with them will go chloride ions. As a result, there will be a high concentration of chloride and hydrogen ions on the gelatin phase side. Eventually, an equilibrium will be established and hydrogen chloride will enter the gelatin phase at the same rate as it leaves it to pass into the solution.

In order to maintain electrical neutrality the hydrogen chloride must go through the membrane as a molecule rather than as a free chloride ion or hydrogen ion, i.e., each chloride ion that goes through the membrane must carry with it a hydrogen ion so that the same charge relationship can be maintained on both sides of the membrane.

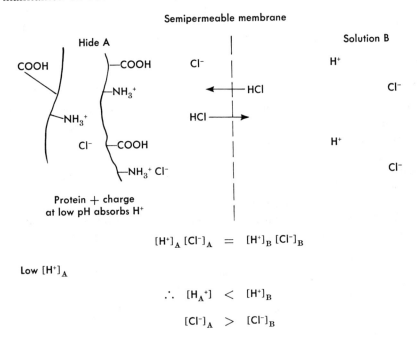

Semipermeable membrane

Hide A

Solution B

COOH

—COOH   Cl⁻   H⁺

—NH₃⁺   ◄—HCl   Cl⁻

—NH₃⁺   HCl—►

Cl⁻ —COOH   H⁺

—NH₃⁺ Cl⁻   Cl⁻

Protein + charge
at low pH absorbs H⁺

$$[H^+]_A [Cl^-]_A = [H^+]_B [Cl^-]_B$$

Low $[H^+]_A$

$$\therefore [H_A^+] < [H^+]_B$$

$$[Cl^-]_A > [Cl^-]_B$$

In accordance with the law of mass action then, the concentration of the hydrogen ion times the concentration of the chloride ion on the gelatin side must equal the concentration of the hydrogen ion times the concentration of the chloride ion on the water phase side. The hydrogen ions on the gelatin side, however, are present in much lower concentration than the chloride ions since the hydrogen ions have been fixed by the hide protein. Therefore, the concentration of chloride ions on the gelatin side must be much higher than that of the chloride ion on the water phase side. If we started out with pure gelatin and a hydrochloric acid solution, the concentration of hydrogen ion and chloride ion on the outside must be equal. At equilibrium, therefore, the concentration of chloride ion in the gelatin phase is much higher than that of chloride ion in the water phase, and an osmotic pressure difference will result.

With the semipermeable membrane, on the gelatin side there will be a strong drawing of the water into the cells or into the gelatin block, and this osmotic pressure can be calculated. In accordance with the laws of osmotic pressure, the pressure is equal to RT* times the concentration of the excess diffusible ion. At the point where the acid of the protein (collagen) is completely bound, the osmotic pressure can reach values as high as 450 atm, or 6000 psi. It is a small wonder that the addition of strong acids to a hide will result in swelling and mechanical damage to the hide fiber. Swelling can be considered to be an initial step toward the complete solubilization by hydrolysis of the hide protein.

The same calculations can be conducted on the basic side for the absorption of the hydroxyl ion with the addition of NaOH or any other base. The swelling again follows the base-binding capacity of the protein, and above the isoelectric point the relative volume of the swollen fiber will be increased in accordance with the alkali bound.

$$[H^+]_A [Cl^-]_A \;=\; [H^+]_B [Cl]_B$$

$$[H^+]_A \;<\; [H^+]_B$$

$$[Cl^-]_A \;>\; [Cl^-]_B$$

But $[Cl^-]_B$ is high due to presence of NaCl, therefore the concentration difference is small and the osmotic pressure is small.

*R is the gas constant.
T is the absolute temperature.

If we consider the same system, i.e., the fixation of hydrochloric acid by the hide-gelatin system, and add to it sodium chloride, the calculations take a different form. Hydrochloric acid again enters into the hide; hydrogen ion is fixed by the hide protein, and chloride ion remains with the hide protein, forming the gelatin chloride. The concentration of hydrogen chloride in the water phase is then decreased. Sodium chloride also diffuses across the membrane, but to a lesser extent since it is not bound to as great an extent by the hide protein as is the hydrogen ion. At equilibrium, therefore, the diffusion of chloride ion across the membrane can be conducted either with accompanying hydrogen ion as chloride, or diffused back and forth as sodium chloride by carrying a sodium ion. Under these circumstances the hydrogen ion concentration in the solution in the gelatin phase may be lower than the hydrogen ion concentration in the water phase. The chloride ion concentration, on the other hand, could be higher on the outside than on the inside, and as a result, the concentration of ions in the water phase may not be greatly different from that of ions in the gelatin phase. Thus, there is considerably less osmotic effect in the presence of salt.

The control of swelling by salt is of great importance in leather manufacture. For example, in pickling NaCl is used in solution and the osmotic effects of swelling are greatly decreased so that no damage to the skin can occur. The greater the quantity of the salt, the flatter will be the hide and the less acid swelling effects. The skillful balance of salt and acid during the tanning process determines the condition of the fiber at the time of tannage. The smoothness of grain and fullness of tannage are, to a great extent, dependent upon these factors.

In the liming operation the alkaline swelling under the buffered system with the calcium hydroxide permits the partial hydrolysis or opening up of the fibers to impart desirable qualities into the leather.

The result of these varying osmotic forces as a function of pH can be expressed in the accompanying swelling curve. This curve is typical for all raw hide and gelatin and shows the minimum swelling at the isoelectric point and increased swelling at both the acid and alkaline ranges. On the acid side the swelling gradually increases until a pH value of about 3 is obtained. At pH 3 essentially all the carboxylic groups of the protein are ionized and the maximum osmotic forces result. Below pH 3 the additional acids necessary to decrease the pH are no longer absorbed by the hide, but rather remain out in the solution and have a repressive action on the osmotic swelling in the same manner that additional salts might have. The curve, therefore, drops down at these higher concentrations. In practical tanning very strong acid conditions in the absence of salt are so deleterious to the hide, and the swelling is so excessive, that the dropping off of the curve at a very low pH value is seldom observed.

Above the isoelectric point, up to a pH of 10, there is very little swelling of the hide protein. This is the range of the pH of the bated skin. Above pH 10 and up to pH 12 the swelling increases with the absorption of alkali, as indicated. Hides in the limed condition are in the area of maximum swelling, and the degree of swelling will depend upon the time in the alkali for the even distribution of the lime and upon the gradual chemical changes in

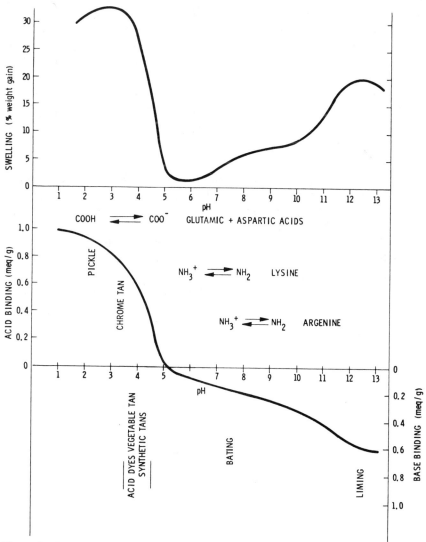

Figure 3 Relationship between pH, swelling, and chemical reactivity.

the hide protein. Therefore, in long liming periods the swelling will continue and the hides will continue to gain weight. Upon deliming and bating, the pH is brought down and the swelling of the skin falls drastically to the non-swelling range between pH 9 and pH 6. A cut on the skin with a knife and the addition of an indicator (phenolphthalein or bromothymol blue), will indicate the "lime streak" in measuring the pH in the center of the skin. During the bating of side leather and sole leather there will be lime streak, indicating high pH in the center of the skin and a differential swelling between the grain and the center layers. The differential swelling which takes place here will cause some distortion of the grain fibers and there will be an internal stress set up at the point where the pH changes from a non-swelling to a swelling area.

The control of the forces of differential swelling can be used by the tanner to control leather quality and grain character. Shrunken grain effects, for example, are often based on such processes.

LOOKING AHEAD

The study of protein chemistry is being actively pursued by research workers in the biochemical, medical, and industrial fields. The results of this research are being used for practical purposes by leather technologists. The reconstitution of hide fiber from scrap hide pieces into a leather type material is but one practical application of such research.

The leather technologist should try to obtain a fundamental understanding of protein chemistry not only for a better understanding of his present processes but also to better evaluate the practicality of new systems.

REFERENCES

Bull, H. B., "Biochemistry of the Lipids," New York, John Wiley & Sons, 1937.

Cohen, E. J., and Elsall, J. T., "Protein Amino Acids and Peptides," New York, Reinhold, 1943.

Fox, S. W., and Foster, J. F., "Protein Chemistry," New York, John Wiley & Sons, 1957.

Gustavson, K. H., "The Chemistry and Reactivity of Collagen," New York, Academic Press, 1956.

Gustavson, K. H., "The Chemistry of the Tanning Process," New York, Academic Press, 1956.

Highberger, J. H., "The Chemistry and Technology of Leather" (O'Flaherty, Roddy, and Lollar, ed.), Vol. 1, p. 65, New York, Reinhold, 1956.

Hilditch, T. P., "The Chemical Constitution of Natural Fats," New York, John Wiley & Sons, 1941.

Koppenhoefer, R. M., "The Chemistry and Technology of Leather" (O'Flaherty, Roddy, and Lollar, ed.), Vol. 1, p. 41, New York, Reinhold, 1956.

Lewis, W. K., Squires, L., and Broughton, G. "Industrial Chemistry of Colloidal and Amorphous Materials," New York, Macmillan, 1942.

McLaughlin, G. D., and Theis, E. R., "The Chemistry of Leather Manufacture," ACS Monograph 101, New York, Reinhold, 1945.

Pauling, L., and Corey, R. P., *Proc. Nat. Acad. Sci. U.S.A.,* **37**, 241 (1951).

TANNERY OPERATION

5

Economic Considerations

No discussion of the technology of an industry would be complete without consideration of the basic economic principles upon which the industry operates. These principles are the basis upon which technical decisions must be made. It is also necessary that we consider the direction in which the industry is going and the forces which are affecting the technological problems. The raw materials for the leather industry are the hides and skins which are incidental by-products of the meat industry. Hides and skins are basic commodities on a world market. International competition occurs for any skin or hide, regardless of where it is produced. The tanner is faced with a material supply that is independent of the demand for leather in his own area.

In the United States, population growth has increased the demand for shoes and other leather goods. The per capita consumption of shoes has remained essentially constant regardless of the expansion in the standard of living. This is true not only of shoes, but of most other soft goods, and this has remained a source of great concern not only for the leather industry but for all the soft goods industries. The demand of the American public for leather goods has not kept pace with its demand for beef. Leather substitutes have also taken an increased proportion of the American leather market.

In shoe upper leather, the population increase and the expansion in shoe production has resulted in a general increase in the volume of leather being used. The ultimate level of sole leather demand, however, has reached a constant value with an essentially constant percentage of shoes with leather soles being produced. The demand for leathers other than cattlehide for linings, garments, handbags, etc., also follows population growth, and with this demand offsetting the losses due to the sole leather industry, the net leather production in the United States has gradually decreased with an increase in production in some other countries.

TABLE 1

Leather Usage Domestic Production

Item	Million Ft.^{2}	%
Shoes	625	56
Garments	100	9
Work Shoes	111	10
Handbags	70	6.5
Leather Goods	70	6.5
Belts	45	4
Miscellaneous	90	8

BREAKDOWN OF LEATHER PRODUCTIVITY COSTS

In considering the economic tannery operation, there are several major categories in which these costs fall. These would not be parallel with an accountant's cost calculation, but are significant and perhaps unique to the leather industry. The categories we will discuss individually include raw material, labor, chemicals, machinery and equipment, effluent treatment costs, and marketing.

Raw material cost

Raw material cost, up until recently (ten years or so) was the largest single factor in the production of leather. The value of the raw materials determined, to a great extent, the value of the leather and the cost of the leather to the shoe manufacturer. The manufacturer, or leather buyer, would watch the listed hide prices, and this was a major factor in his negotiation of leather prices with the tanner. This is still true today, but to a much less extent.

The other costs of leather production have increased to a larger percentage of the proportionate sales dollar volume in most types of leather. Although the raw materials are a relatively low percentage in the cattle hide and heavy leathers, they are the dominant factor in the fancy leathers. The added value of leather production in a reptile leather, therefore, is considerably less percentagewise than the added value in a heavy leather. The risk involved in the manufacture of such fancy leathers is high because there is a large investment in the raw stock. The reserve for overhead, contingencies, and profit in fancy leather is, by necessity, higher because of the risks involved.

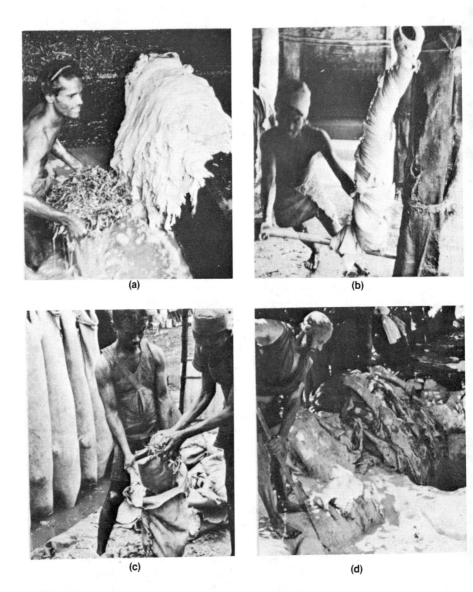

(a) (b)

(c) (d)

Figure 1 Although modern leather production is highly mechanized and is under close chemical control, primitive methods are still in use in some parts of the world. **(a)** East India Tannage. **(b)** East India Tannage. Large quantities of skins have been exported by India based on these methods. The skins are retanned and finished in Europe and United States to make a wide variety of fine leathers. The East India tannage is gradually being replaced by very modern methods. **(c)** Indian bag tannage for sandal leather for local use and export. **(d)** Pit lining in Afghanistan for local leather. *(Author's Photographs)*

Both the availability of hides and their price are somewhat artificial. As mentioned before, the United States is a hide-exporting nation. The hides are bought in large quantities by two major customers: Japan and the Eastern European socialist block. Therefore, on occasion, relatively high costs of hide may be found in the United States and Europe, while the demand for leather remains relatively low. Some nations have placed restrictions on the export of their hides and instead are producing leather for sale on the world market. This has increased the demand for hides from the United States, forcing the price quite high for the American tanners. As hide prices are forced high, they can conversely drop when the demand abroad is suddenly slack.

Labor costs

Labor costs in the leather industry are rising on a man-hour basis. The leather industry, in trying to obtain help for its wide variety of semiskilled and unskilled operations, must pay the going market price. Even though a tanner may be willing to pay the necessary wages to get competent help, the shortage of labor in many areas has forced a drastic decrease in the total labor force in the tannery. Productivity per man-hour has undergone tremendous increases as a result of increased use of automation. Laborsaving devices are being employed so that the man-hour productivity, by the use of better equipment, larger capacity machines, automatic handling equipment, stackers, etc. has, during the last twenty-five years, increased fourfold the man-hour productivity on side leather.

While fluctuations in the market, particularly at today's cost of money, are greatly increased, the increased productivity of labor due to mechanization has allowed American tanners to compete on equal basis with many other nations having a much lower hourly wage. During the late seventies, the cost of labor in Europe and in Japan increased dramatically, and the efficiency of production did not accompany the increase in hourly rates of wages. As a result, there has been a marked increase in the importing of leather by Europe from the developing countries as well as the United States.

Chemicals

The ingenuity of leather chemists has developed new chemicals for treatment of leather, allowing much better characteristics to be built into the leather than were dreamed of a generation ago. New fatliquors and syntans as well as silicones and other water repellent treatments are giving soft, supple, durable leathers that were impossible previously. These new systems have also helped decrease labor costs and speed production. The proportionated amount of leather production costs going into leather chemical specialties is increasing annually and has been doing so for the last few decades.

Machinery and equipment

The need to save on labor has been a prime factor in the increased use of machinery by the tanning industry. In addition to automation and the resultant decrease in labor costs, equipment is being built which will aid in producing specific effects in leather. In this category, specialized drying equipment, for example, has permitted drying at lower temperatures and higher rates than ever before and will make possible new specific qualities in the leather. Finishing machinery such as flow coaters and automatic spray equipment are not only laborsaving devices but also permit the accurate and efficient application of chemicals.

One of the problems of the industry is obsolescence. Many of the older tanneries are using machines which were installed a generation or more ago. The cost of operating a plant of this type may be high, but the machinery has long since been written off and the plant is being operated simply because it is there. As the maintenance cost on these machines becomes higher, the plant becomes more marginal in its productive efficiency and, eventually, such factories are forced to close in favor of the more modern, more efficient, and usually better-financed operations.

The consolidation of tanneries into fewer, larger units has also been accompanied by changes in machinery. Beamhouse equipment is a case in point. During the past decade, unhairing bate pickle and tan processes have been developed into multistep operations without removing the hides from the drum or hide processor.

The average batch size in new equipment is about twice that of the older system.

Effluent treatment

One of the major considerations in the design and operation of a tannery and development of tannery processes is the consideration of effluent. Tanneries result in large amounts of waste solids and chemicals. Leather chemists' efforts during the seventies have been dominated by effluent considerations. More time and dollars have been spent on research and development in the effluent field for the tanneries than on improvement in the quality of leather or its cost of production. Most companies that are paying for effluent treatment plants and their development cost have increased operating expenses and must have increased revenue to compensate for this. The cost of effluent treatment is approximately the same whether it has been done by the tannery as a primary discharge, or whether the tanners discharge into a municipal system.

A lesser emphasis on environmental considerations gives an additional advantage to the developing nations. This, of course, differs greatly in different leather-producing areas.

Basis of profitability in tannery operation

An old-time tanner friend once said, "It's very easy to make money in the leather industry. All you have to do is buy your hides cheap, keep your labor costs in line, and produce a good-quality product that will command a good price." This profound statement, of course, is true of any industry, but there are several means of attaining this happy state.

Shrewd purchasing of raw materials has been the key to profit in the leather industry for many years, but it is only recently that tanners have recognized that speculation on raw stock is not the answer to making money in a tanning operation. If a tanner can determine the fluctuations of raw stock prices accurately, he is much better off not being in the tanning industry at all but would do better buying and selling hides on speculation. He could, therefore, eliminate his investment in machinery and equipment and eliminate the problems of selling the finished leather. Raw stock speculation has become less and less of a factor in the production of modern leathers. The tanners are now relying more on their individual abilities to give added value to their product.

The question of added profits can take two forms: high quality or low cost; very seldom are both obtainable at once. If a tanner operates a very large tannery he must, by necessity, turn out a large volume of goods in order to keep the plant operating efficiently. For sales on a large volume, the goods being sold cannot be produced in small lots, nor can they be high fashion items. The basic staple materials of standard colors must be produced in large quantities at low prices. Price is the tanner's only real advantage in obtaining the necessary large portion of the market that he needs to keep his plant operating profitably. A large producer, therefore, will operate in long, steady runs on the same type of leather. He will keep the factory running steadily with little overtime payment and will be very price-conscious in the purchase of chemicals and very productivity-conscious in considering labor costs and equipment.

The small producer cannot hope to attain the same degree of efficiency as the large-volume producer with his long, steady lines of production and high degree of automation. In his sales the small producer needs a relatively smaller portion of the total market, and in order to obtain this, he must offer some factors of quality not obtainable from the large producer. The smaller producer maintains, in general, closer contact with the matters of style in color, grain character, and possibly service to his customers.

Contract tanning

The separation of raw stock speculation and leather sales from the skills of operating an efficient tannery creates the position of the contract tanner. Contract tanning refers to the practice of a leather merchant buying hides and contracting with the tanner to produce a particular leather. The leather merchant applies his skills as a seller of leather and a purchaser of raw stock, while the tanner devotes his attention to managing the tannery and producing quality leather. This division of capitalization between the two principals has certain advantages and disadvantages. For the tanner, the total amount of capital necessary is decreased in that he does not own the hides which are being worked through his tannery. An additional advantage to the tanner is that he is not concerned with the sale of the leather, and he is working against a known fee for his services of operating the tannery. The disadvantage to the tanner is that he does not share in the profits resulting from extra quality in production or unique characteristics of the leather. The leather merchant engaged in contract tanning has the advantage of being able to deal in hides and skins purchased against a sales contract, rather than having his capital tied up in the physical tannery. The amount of capital needed by the leather merchant, therefore, is greatly decreased and he is free to apply his skills in the marketplace without the worry and problems of technical production. The disadvantage to the leather merchant, of course, is that he cannot develop, for his own use, an exclusive line or quality of leathers to give him an advantage over his competitors.

All the above factors—the cost of machinery, the cost of labor, the cost of chemicals, the volume sales, policies of the company, the capital reserves of the company, the depreciation of the plant, and whether the tanner is a contract tanner or a merchant tanner—affect the technological decisions with regard to the acceptance of new processes, new machinery, or investments in technical developments and research.

The added risk and the added cost of inventory make the merchant tanner's position much more expensive. A merchant tanner must, therefore, be very well financed. A contract tanner, on the other hand, operates a tannery without concern for the cost of raw materials or the salability of the leather. The hides are sent to him to make leather to a specification for a certain cost. As long as these specifications are met, and costs are competitive, the contract tanner should be successful.

In a tannery the inventory, in general, exceeds the value of the machinery. The machinery is less valuable than the real estate. It is, therefore, much easier to enter the contract tanning business in a leased building than it is to be a merchant tanner under any circumstances. Many people in the industry have become wealthy, either as merchant tanners or as contract tanners. A like number of people have suffered great financial losses both as merchant tanners and contract tanners. You can win or lose, either way.

Tannery location

There are three general areas of leather production in the United States which have been established by the needs of the industry in the past. On the Eastern seaboard, from Maine to Delaware, tanneries have been located to supply the shoe and garment industries for eastern manufacturers. The proximity of these areas to good seaports has led to particular emphasis on the light leathers based on imported hides. Kid leather production and the sheepskin garment industry have been concentrated on the eastern seaboard states. There have been developed certain areas of specialization, such as the kidskins in the Philadelphia–Wilmington area; reptile and fancy leathers in the Newark area; glove leathers in the Gloversville, New York area; garment leathers of sheepskin suedes and cape leathers in New England. In recent years, with the shifts from imported to domestic hides, large quantities of shoe upper leather are being produced, particularly in New England. The development of the shoe upper leather industry in New England has spread more towards Maine and New Hampshire in search of low labor costs.

The heavy leather industry for sole leathers and bag, case and strap leathers, was originally established in this country in the areas where the production of the natural vegetable tannin materials was possible. With the availability of the chestnut trees, sole leather tanneries were established in Virginia, West Virginia, Tennessee, and Pennsylvania. The chestnut trees have long since gone but the sole leather industry still remains concentrated in these areas.

The availability of cattle hides in the Chicago-Milwaukee area resulted in the establishment of shoe upper leather tanneries in these cities. In the Midwest, Wisconsin and Illinois, particularly, stand as the major centers of shoe upper leather production in the United States.

With the recent trends toward greater production of leather near hide sources, the earlier stages of leather production are being shifted to the Midwest, particularly to Iowa, Nebraska, Missouri, and Texas. The highly mechanized finishing and coloring operations remain closer to the shoe production centers.

The expansion of the leather industries, particularly in developing nations, has decreased the availability of goatskins and many other types of imported skins to the American market. As a result, there has been a tremendous decrease in the number of tanneries operating in the Philadelphia-Wilmington area. Those that remain are operating primarily on cattle hides.

Practical Tanning Procedures

The purpose of this section is to orient the reader into the general methods of production as applied in several different types of leather. For each of

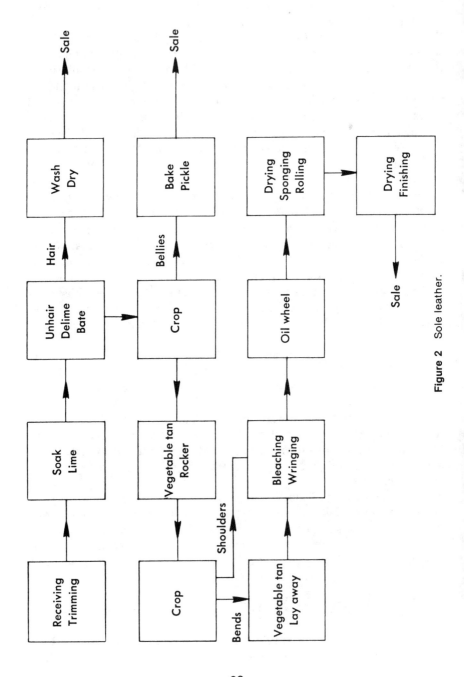

Figure 2 Sole leather.

these leathers the production is dependent upon economic and technical considerations. This is due to the type of raw materials being used and also to the end use of the leather. Three separate significant types of leather are used as examples: (1) sole leather as an example of heavy leather production by the vegetable tanning process; (2) side upper leather which is a high speed production oriented process based on chrome tanning; (3) suede garment leather based on New Zealand or domestic pickled skins. The methods outlined are not intended to be recipes for practical leather production nor are they necessarily the latest developments of the art. Rather, they are to be taken as the typical present state of the industry. The more detailed discussions of processes in the subsequent chapters will cover these newer developments.

HEAVY LEATHER—SOLE LEATHER

Heavy leathers and sole leathers are produced by the vegetable tanning process, the oldest of any process in use in the industry. The method is used in the production of mechanical leathers, bag, case, strap, and sole leather. The hides are of heavy steer type because thickness is a prime consideration. The tightness of the fiber and the angle of weave are of importance in that good hide substance is necessary to the manufacture of durable leather.

Sole leather is sold by the pound, although there is a distinct trend towards pricing the sole leathers by the pair as cut soles. Raw materials for sole leather production are purchased year round by the sole leather tanneries; the main source of supply is the American heavy cattle hide industry.

The processes involved in making sole leather by the vegetable tanning procedure are very long. Consequently, large portions of the tanner's capital are tied up in goods in process. A relatively small inventory of hides is kept on hand for his production needs, since the heavy hides purchased are going to go into heavy leather and there is little difference in requirements of the hides purchased than there is for side leather. The goods in process in sole leather manufacture may be in tanning for as much as two months. Under the older systems of sole leather production the vegetable tannages took considerably longer and it was not uncommon for sole leather tanners to have leather in production for as long as four to six months. Since a decrease in production time results in decrease in inventory and a freeing of capital, there have been strong pressures on the sole leather tanners to shorten process time. Thus, as technological advances were made and the sole leather production time decreased, less equipment was necessary in the form of the rockers and tan yards. As a result, the sole leather tannery decreased in size and parts were cannibalized in order to maintain the remaining production sections of the rocker. Sole leather production has decreased since World War II but is maintaining a fairly constant rate as of the last five years. The demand for sole leather goes primarily into the better quality shoes.

Sole leather is a staple item used in shoe manufacture and is not subject to the great whims of color and texture change that we find in other types of leather. There are fews changes in the industry with regard to making sole leathers much more durable and more water resistant. Most of the changes in the sole leather industry to date have been toward speeding production so as to free more capital. These are not necessarily advancements in leather quality for performance characteristics.

PRODUCTION OF SOLE LEATHER

A typical sole leather tannery in the United States would produce approximately 2,000 hides a day. Hides processed are heavy native steers or Colorado steers in 70 pound and up classifications. The hides may be received either brine cured, or conventionally salt cured.

The hides are trimmed and soaked. Following the soaking, which is usually an overnight process, the hides will be fleshed. In the case of prefleshed hides this step will be eliminated. Hides are then placed in lime for unhairing. In sole leather production it is necessary for the grain to be clean and the hair follicle to be free of hair roots. To accomplish this long liming procedure, hair-saving systems are generally used. The normal procedure for sole leather liming is to use a series of pits. The hides are linked together by small strands of rope into a long chain. The first pit contains lime liquors with sharpening agents such as sodium sulfhydrate and dimethylamine. The hair loosening process starts in the first bath. The next day the skins are hooked on a mechanical device called a "reel" and lifted over from the first pit into the second pit. The second pit contains less sharpening agents, and as the skins are pulled over they are laid down into the pit without agitation. The third day they are pulled over again and this is repeated each day.

Liming procedures for seven days are quite common for this industry. The hides, after being removed from the last lime, are ready for unhairing. The hair is recovered, washed, dried, and sold. Hair is usually of good quality. The hides may be further scudded (machine scraped) to remove any additional material in the hair follicle. Hides are fleshed in the limed condition and may be cropped.

Cropping may be done in either of several different ways. By cutting off the shoulders and bellies, limed shoulders and also limed bellies are obtained. Bellies may be pickled or vegetable tanned separately or sold in the lime condition. The shoulders are either vegetable tanned separately or they are left on the skins and only the bellies are removed. The choice depends upon the most efficient use of equipment, current economics, and the type of hides being worked. The hides are tied and hung in rocker racks and tanned by the vegetable tanning process.

Prior to vegetable tanning, the hides in the lime condition are put in rockers containing deliming and bating materials. The deliming of heavy cattlehides is to adjust the pH so as to make the skins receptive to the vegetable tanning without having precipitation of the vegetable tannins in or on the skin. Since the hides have been given a long liming, very little bating action is necessary. Deliming proceeds until the skins have gained a desired pH. The skins are then lifted over into the rocker system and tanned by the vegetable tanning process. In the vegetable tanning process by the rocker system, a series of rockers is employed in which the concentration of the tanning materials starts out low and is gradually increased as the tannage proceeds. This is accomplished by moving the hides from one rocker to another or by pumping the liquids. Control of pH is of importance and it is necessary that mild tanning action be employed initially to obtain good smooth grain and not to case harden the surface of the leather. It usually takes three weeks for the tanning material to penetrate through to the center of the skin. Skins are removed, wrung, and may be further cropped. The shoulders are taken off and sold as double rough shoulders or the bellies may be taken off and processed separately for insoles, lining leathers, etc. The most valuable part of the skin is the bend. Double bends may be cut into single bends for ease of handling from this point on.

Heavy bends are placed on vats called layer vats and more extract is added. The bends lie in the layer vats for one or two weeks in a warm vegetable tan liquor, and again the liquor may be strengthened and heated to gain better penetration and fixation of the strong tanning liquors. When the desired quantity of vegetable tanning materials has been taken up, it is placed in a weak tan liquor solution that is slightly acidified. This helps to remove some of the excess tannins on the surface and will aid in the cleaning up of the skin and development of the proper color. The bends are placed into a drum and scrubbed by the passing of solutions into the rotating drum. This will clean off the sediment on the surface.

Vegetable tan liquor is used at 95–100° F and after about a quarter of an hour the vegetable tannins will be even and the skins will be cleaned. The scrubbed side bends are then put into racks and bleached by dips into a solution which then removes some of the surface tannins from the skin. This is a series of quick dips in alkali or acid to clean the surface. After bleaching, the bends are wrung and sent to an oil wheel. In an oil wheel the materials such as lignosulfonate, epsom salts, corn sugar, oils, and specialty chemicals are used. These materials are drummed into the leather, and coprecipitate in the skins. This process usually takes about one hour. The leather is "set out" to smooth and dry. After drying, the skins are wet again, piled down, and sponged on the grain surface with an oil sponging compound. The leather goes through a rolling operation in which the surface of the skin is rolled with a heavy cylinder to compress the fibers. The leather may be sponged, dried, and rolled several times. After the leather has achieved the right degree of resiliency it is sponged with wax coating materials and dried. It is then brushed and dry-rolled to smooth out the grain.

In the manufacture of sole leather most of the time of the leather production is taken up in the vegetable tanning, in the rockers, and in the lay away in the tan vats. Hand labor is confined to the setting, drying, conditioning, and rolling operations. Much more labor is employed in the latter stages of sole leather production then in the earlier stages.

In the selling of vegetable tan leathers the price of the hides is of importance but a major portion of the expense is in the cost of the vegetable tanning materials. For successful operation of a sole leather tannery under present market conditions it is essential that effective use of the vegetable tanning materials be attained and relatively fast tanning process be achieved. In recent times the practice of impregnating or loading sole leathers with epsom salts and oils is beginning to be replaced by the use of impregnating compounds which result in sole leather with better wear and greater water repellency.

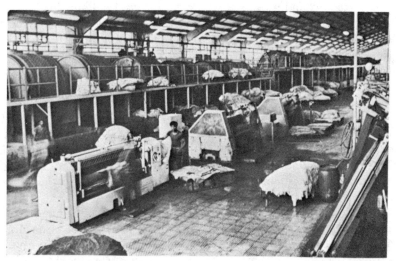

Figure 3 Modern tannery producing wet blue and pickled sheepskins, hair sheep and goatskins—Azar Leather. Melli International Group—Iran. *(Courtesy Vallero, Cesare & Figle.)*

SIDE LEATHER

The manufacturer of shoe upper leather from cattle hides is one of the most dynamic and most competitive sections of the leather industry. There are seasonal fashion changes in color and texture. A variety of leathers can be made from the same raw material. The tanner's skill in management and his artistry in fashions are the major factors in the success of the enterprise.

Raw materials can be purchased regularly in all seasons. The major raw material is the cattle hide from the Western steer beef. In recent times many more of these hides are received prefleshed and brine cured. This is particularly true in heavier classifications. Tanners buy their raw materials with a particular type of leather in mind and base their purchases on the experience gained from this type of hide. The average side leather tannery today, in the United States, can produce about two thousand hides (or four thousand sides) per day. This is the equivalent of about two carloads of hides. The inventory of hides waiting for production is usually about two weeks supply. This will vary considerably depending upon the availability of hides in the local area and the physical facilities at the tannery.

Hides, when received, are trimmed and placed to soak. Some tanners may soak overnight, others wet wheel the hides by placing them in a drum for two to four hours. The soaked hides may be fleshed or may be sent directly into liming. In liming there may be either a hair-saving or a hair-burning process. The hair-saving process will usually require two days in the drums or paddles. The hair-burning process will usually need one day. The limed sides are then refleshed and put into the drums for a bate, pickle, and tan.

The bate, pickle, and tan operation is quite common in the shoe upper leather production in the United States, and in modern installations using large drums, the drums will operate on 10,000–15,000 pounds of limed weight. In the bate, pickle, and tan operation the hides are converted into chrome tanned leather. The process may operate on a 8–24 hour cycle. Chrome tanned stock is piled down for a day, wrung, and graded for thickness desired and for the type of leather to be produced. Leather is then sent through the splitting machine and cut to uniform thickness. Due to the structure differences in the hide, the bend area and back area is split slightly thinner than the belly area. To gain further uniformity of thickness, the sides are shaved to the desired thickness with greater precision than is possible with the splitting machine. Splits separated from the sides are trimmed and made into other types of leather. Although many side leather tanners in the United States also work splits to completion, most tanners today prefer to sell the splits "in the blue" to tanners who work specifically in this field.

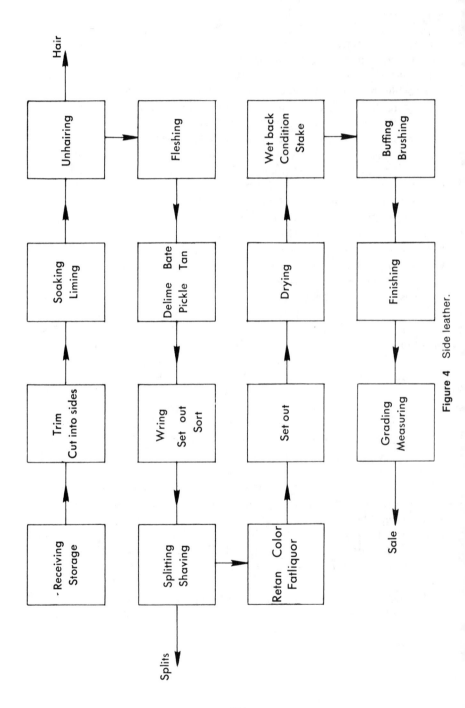

Figure 4   Side leather.

74

The split tanners, upon receiving the splits, trim (possibly re-split) and shave them to a uniform thickness. The splits are retanned, colored, and fatliquored. Splits may be made into "rough buck" type shoe uppers, work glove leathers, heavy garment leathers, insole leathers, or may be given a finish and sold for shoe uppers.

The grain leathers from the shaving machine are weighed into packs for retanning, coloring, and fatliquoring. The use of dyes, syntans, extracts and oil combinations at this point determines the type of leather to be produced. After retanning and fatliquoring, the stock is wrung, set out, and dried.

The leather may be dried by any of four common methods. (1) Hang drying: This is the cheapest and easiest method. The leather is hung on hooks in a loft and is dried by circulation of warm air around the leather; (2) Pasting: The leather is spread out on glass or porcelain plate to gain the maximum area yield and greatest smoothness of the grain. Pasting is by far one of the most popular methods of drying leather in the United States; (3) Toggling: The leather may be toggled, in which process a small drying unit is used and the leather is stretched and clamped on a screen; (4) Vacuum drying may also be used.

Once the leather is dry and the fibers have fixed the soluble materials in the skin, the leather is wet back and allowed to equilibrate at a proper humidity condition for uniformity. The leather at the desired uniform moisture content is staked to soften the fiber by mechanical action.

In order to remove grain imperfections, the leather is lightly sandpapered prior to finishing. Buffing will not be used on those hides having excellent grain surface. This leather will be sold as full grain leather. Full grain leather gets premium price and is much more pleasing to the eye than the corrected grain (buffed) leathers. Full grain leathers can take clear top finishes whereas corrected grain leathers are usually treated with more heavily pigmented finishes.

Finishing operations may require three to five coats, and impregnation may also be desired. Pigment and base coat finishes are usually emulsion systems; the top coat may be a lacquer and may contain some waxes. An embossing press is usually employed to smoothen the finish and in some cases to imprint a pattern into the surface of the leather. If the leather is to be embossed into a pattern, usually vegetable extract is used in the retannage. Upon completing of the finishing operation the leather is sorted for grades, measured, bundled, and shipped. It is common practice in the industry to bundle leather in paper before shipping to the shoe factory.

In the manufacture of side leather the labor is by far the most important factor of cost. Through the chrome tanning operation, most tanneries operate on approximately the same cost level. Beyond this point a wide variety of chemicals and procedures is used and the cost of materials,

Figure 5   Modern tannery for the production of pigskins—Industria Usnja Vrhnika—Jugo-slavija—*(Courtesy Vallero, Cesare & Figle.)*

machinery, and labor employed may vary greatly from one tannery to another. The amount of labor and types of treatments being used will depend upon the leather being produced.

The side upper leather tanner is constantly plagued with fashion changes in texture, colors, and finish, and he is constantly striving to develop a unique line of leathers for his customers. Practical experimentation in side upper leather tannery is concentrated primarily in two areas: (1) the retan, coloring, and fatliquoring operation, and (2) the finishing operation. In both these areas the new leathers are developed. Once a good basic formulation for unhairing, bate, pickle, and tan is found, it is usually kept quite consistent over a period of years. The quality of leather produced for a given hide may vary greatly depending upon the quality control and the tannage. The ultimate value may vary depending upon the type of leather produced.

In the manufacture of side leather the profitability of the company is based upon the gain of the maximum economic advantage through proper selection of the hides for the intended purpose; the hide buyer must buy the lowest priced hide possible that is of sufficient quality to make the desired leather. Consistent quality of production through close chemical control is also needed. A good choice of equipment and automation to reduce labor costs and an eye for style and beauty all play a role in the successful operation of a shoe upper leather tannery.

The manufacture of side leather represents a very large capital investment in plant and equipment as well as a very large hide inventory. In order that the reader be oriented in relative costs the following data are presented. In these data the costs of equipment and buildings are given at today's estimated new replacement costs. The data are approximate, based on discussions with tanners and may not be typical of any particular plant. The type of leather produced and local factors may greatly effect the costs.

The following table assumes a side leather tannery (merchant tanner) producing 1,000 hides (2,000 sides per day) making an average side upper leather from big packer hides.

THE MANUFACTURE OF GARMENT SUEDE LEATHER

In the manufacture of garment suede leather the raw materials are primarily the pickled sheep skins of New Zealand lamb and sheep. Since the materials come from New Zealand on a seasonal basis, they must be purchased well in advance in order to assure a good supply of raw materials for quality production during the season to come. Purchase of New Zealand sheep skins and pickled sheep skins is highly speculative and very competitive. The tannery must buy the skins from the dealer while they are still in New Zealand. The skins are stored in large wooden casks or wrapped in plastic. The pro-

TABLE 2

Breakdown of Employees and Inventory by Departments Based on 2,000 Hides per Day

Department	No. of Employees	Hides	Value per Hide	Total Value
Hides on Order	0	20,000	$55.00	$1,100,000
Receiving	10	14,000	59.85	837,900
Blueing	32	8,000	69.75	558,000
Sort/Split/Shave	8	6,000	64.25	385,500
Color, Fatliquor	20	6,000	77.45	464,700
Dry & Finish, Ship	50	40,000	92.85	3,714,000
Administration	20			
Accounts Receivable		40,000	92.85	3,714,000
TOTALS	140			$10,774,100

duction of the skins is worked over a period of months. Since the stock is pickled, no beamhouse is necessary and the production from pickled sheep skins is greatly simplified. Skins, however, do have a large amount of grease and this adds some complication to the production. One of the main factors facing the garment suede manufacturer is the problem of maintaining good quality fiber and strength while meeting the customers' demand for high fashion. There is little quality difference between the skins produced by one tanner and another from the same raw material. The sale is therefore in a highly competitive price-based market. Consistent quality of production is an advantage for some manufacturers. A normal garment suede tannery will work in the neighborhood of 3,600–12,000 skins per day. Skins, when they are received, are inspected using a light to show pinholes and heavy buckiness. The skins are placed in the drum and degreased by the application of solvents and detergents. The degreasing operation is run until the temperature is raised to about 95° F in the drum. The solvent and the detergent are then washed out by the use of brine, and after subsequent brine washes, the skins are free enough of solvent and residual grease so that the chrome tanning may be conducted. The degreasing brine solvent solution is recovered and the solvent distilled, the grease sold, and the solvent reused.

Chrome tanning is conducted in a normal manner in a drum, usually to a full tannage. After the completion of the chrome tannage the skins are piled down for a day or so, wrung, and fleshed on an abrasive wheel. Prior to the fleshing operation the skins may be tumbled with pumice to aid in the fleshing. Chrome tanned fleshed skins are then re-tanned and fatliquored.

Re-tanning operation is simple relative to that which is used in side leather. Since the skins have an open fibered nature, the extracts penetrate very easily. Usually less than 10% powdered extract based on the chrome tanned weight of the skin is employed. Some syntans of the naphthalene sulfonic type may also be used. After the vegetable tans and syntans have fixed, the fatliquoring oils are added. Fatliquored skins are removed from the drum, piled down, set out, and dried. The piled skins are then staked and buffed on the flesh side to produce the suede nap. The buffing operation is conducted on an abrasive wheel. The development of the suede is very much a matter of the operator's skill to develop the skins properly to the desired texture of the nap.

Skins in the crust may be held until a desired color pack is made up. They are usually graded in the crust to determine the type of leather and the color they are to become. Depending upon the depth of shade, the quantities of dye used are up to 15% or more based upon the crusted weight of the leather. One of the main factors in the successful operation of a garment suede tannery is the application of dyes to develop the desired colors, using the least dye at the least cost. Dyes should be even and deep penetrating colors

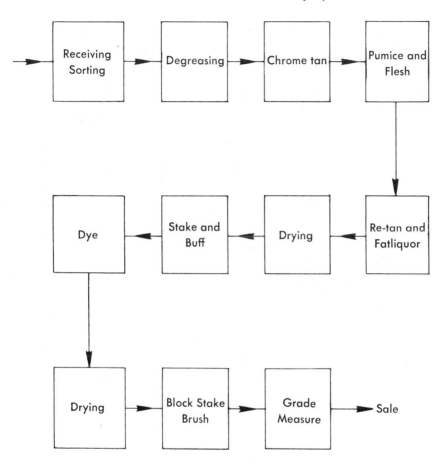

and should not crock or rub off. After dyeing, the skins are set out and dried. They are then trimmed, staked, and block staked to develop the nap further. After block staking they may be polished on a plush wheel to smooth the nap and clear the stock of loosely held fibers. Garment suedes are measured for area and sold on a square foot basis.

Since sheepskin production involves a large number of small skins, its production does not adapt to automation as well as that of shoe upper leather. The sueding and staking operations are done on an individual skin basis. The wet operations of degreasing, tanning, re-tanning, and dyeing a large number of skins are done in a single pack, so the major labor factors are in the drying operations and the dry operations. Sheep skin tanneries are relatively smaller and more simple than side upper tanneries in the amount of both equipment and space that is needed.

Figure 7   Block staking of sheepskin garment suede. This is one of the final steps in sheepskin suede production. It aids in the development of the desired nap. The top unit contains an abrasive paper; the bottom has brushes that shape the skin against the block. *(Courtesy L. H. Hamel Leather Co.)*

GLOVE LEATHERS AND FANCY LEATHERS

Fancy leathers and leathers that are used for gloves are made in very small tanneries from very high priced raw materials. Each of these leathers is quite different, depending on the skins used. The processes involved may be closely related to general processes employed in the shoe upper leather or garment leather business. In light leather production the cost of labor and the cost of materials is fairly high on a per skin basis, but with glove and fancy leathers the value of the raw material is a very high percentage of the final value of the leather. Effective operation of a glove leather or fancy leather tannery is therefore based primarily on the proper purchasing of raw materials, and secondly, on the development from the purchased materials of distinctive leathers of highest quality.

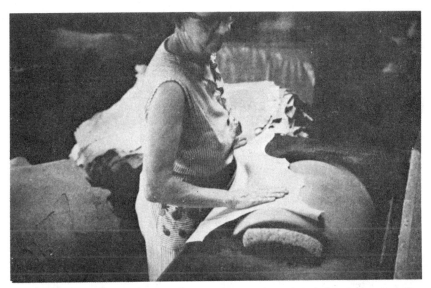

Figure 8   The last step in the manufacture of garment suede is to brush the nap with a plush wheel. This brushing decreases crock and develops a fine nap. *(Courtesy L. H. Hamel Leather Co.)*

Cost of chemicals and the automation of equipment is of less importance in the production of glove and fancy leathers. Glove leather tanneries and fancy leather tanneries, therefore, may look quite small and primitive in comparison to those of shoe upper leather.

The design of a tannery and the design of a production process are based on the raw materials, the end use of the leather, and the properties desired. In considering the use of machinery, labor, and materials, a balance must be made on an economic basis. Each of these items, weighed one against another, must be taken into consideration when judging a tannery and its processes. What is effective and worthwhile in one type of production may not be applicable at all in another. The leather technician should use caution so as not to be fooled by outward appearances; he should make a careful economic evaluation of every technical process. The soundness of technical processes is of little value if they are economically unfeasible.

LOOKING AHEAD

In looking ahead to the trends of the leather industry we can expect more emphasis to be placed on laborsaving approaches, particularly in the side upper leather industry.

Table 3

Value Per Hide Through the Tannery

		$ Value per Hide	$ Value per Foot
Fresh Hides	−80 lb. × $0.6625	$ 53.00	$1.20
Curing Cost		3.00	
Cured Hide	−70 lb. × $.80	56.00	1.27
Fleshed Hide	−55 lb. × $1.02	56.00	1.27
FREIGHT:			
Cured	70 × $.07 = $4.90	60.90	1.38
Fleshed	55 × $.07 = $3.85	59.85	1.36
Assume Fleshed Hide ($59.85)			
Blueing Cost $.18/lb.	$9.90	69.75	1.59
Sort/Split/Shave	2.50	72.25	1.64
Split Credit 20 lbs @ $.40 =	$8.00	64.25	1.46
Color, Fatliquor @ $.30/ft		77.45	1.76
Dry and Finish @ $.35/ft		92.85	2.11
Sales & Administration @ $.25/ft.		103.85	2.36
Profit and Contingencies @ $.14/ft.		110.00	2.50

Complete changes are under way in the manufacture of sole leather, and products with entirely different standards of quality are presently emerging. In the garment suede and light leather industry the changes will be less those of visual differences than those in the area of obtaining greater durability of the leathers and greater ease in dry cleaning, so as to improve the utility of the product.

There are two classes of technological advances in the leather industry: (1) those things which will improve the efficiency of production while keeping quality the same, and (2) those systems which will improve the quality and saleability of the leather and maintain approximately the same cost. Of these two, the most significant and most important for the advancement of the industry is the second.

REFERENCES

"Dictionary of Leather Terminology," New York, Tanners Council of America, 1957.

Freudenberg, W., "International Dictionary of the Leather and Allied Trades," Berlin, Springer-Verlag, 1951.

Glozic, B., "Kozorstov," Zagreb, Tecicka Kniviga, 1954.

Gustavson, K. H., "The Chemistry of the Tanning Process," New York, Academic Press, 1956.

"How American Shoes Are Made," Boston, United Shoe Machinery Corp., 1966.

J. Am. Leather Chemists' Assoc., Supplement No. 11 (1965), Report of the Symposium on Mechanization in the Tannery—June 17, 1964.

McLaughlin, G. D., and Theis, E. R., "The Chemistry of Leather Manufacture," ACS Monograph 101, New York, Reinhold, 1945.

Orthman, A. C., "Tanning Processes," Chicago, Hide and Leather Publishing Co., 1945.

Reed, R., M. Sc., "Science for Students of Leather Technology," London, Pergamon Press, 1966.

Sharphouse, J. H., B. Sc., "The Leather Workers Handbook," London, Vernon Lock, 1963.

Turley, H. G., J. Am. Leather Chemists' Assoc., 57, 500 (1962).

Wilson, J. A., "Modern Practice in Leather Manufacture," New York, Reinhold, 1941.

BEAMHOUSE

6

The term "beamhouse" refers to the processes in the tannery between the removal of the skins or hides from storage and their preparation for tanning. This includes soaking, trimming, fleshing, unhairing, liming, and bating. The term dates back to the time when the hair was removed from the skins by means of a hand beam, i.e., on a sloping, curved table or large log using a two-handled knife. This working of the skins on the beam is of ancient origin and is still in use today. Even in the most modern and sophisticated facilities, some hand work on the beam is needed occasionally for quality improvement.

The beamhouse operation has the distinction of being the most disagreeable step in leather manufacture. It involves the use of bad smelling concoctions which have been responsible for much of the poor name of the leather industry in its community relations. The beamhouse operations are also of tremendous importance in the ultimate quality of the leather. Indeed, in the opinion of most practical tanners, "leather is made in the beamhouse." Beamhouse operations also employ complex principles of biochemistry and inorganic chemistry and are the most difficult areas of leather manufacture for the uninitiated to understand. Practical beamhouse operation, however, can be reduced to relatively simple steps, and good quality leather can be made by close attention to detail and empirical observation of cause and effect.

Liming has been the standard method of unhairing for centuries, and even today the commercial use of any other system is a rarity. We will, then, consider unhairing first as a liming system, and later as non-lime chemical and enzymatic systems.

There are three major steps involved in beamhouse operation: soaking, unhairing, and bating. These steps are common to all types of leather produced, although there are significant variations depending upon the type of skin used and the type of leather to be made. All three steps are closely related and interdependent.

SOAKING

Hides and skins, when received by the tannery, are usually in a condition of preservation based on dehydration. These skins may be dried, as is the case with most goatskins and some tropical cattle hides, or they may be dehydrated by means of salt, either in the form of brine or salt pack cures. The skins must be wet back and brought to a flaccid condition for subsequent operations.

Figure 1 These pictures show leather production as it has been done for thousands of years in the underdeveloped countries. The hides are soaked in lime pits to loosen the hair. Unhairing is done by scraping. The deliming is done with acids formed by fermentation of weak tannin solutions. Vegetable tanning is done in pits using the ancient layering technique. All work is done by simple hand tools. (Author's photographs)

85

Prior to putting the skins in process, they are generally trimmed for the removal of the shanks, tails, and other small appendages that may interfere with the proper operation of machines. Cattle hides may be cropped or sided, depending upon the types of leather to be produced. For side leather production, the skins are generally cut down the backbone, producing two sides. These sides are of a convenient size for machine operation and can be handled by one man. Sole leather and upholstery leather hides will be kept in the whole hide pattern until later in the production.

Soaking is generally accomplished by placing the skins in water, which may contain an additive. In the case of green-salted cattle hides, the hides are placed in a drum and cold water is introduced. The water will dissolve the curing salts and decrease the concentration of salt around the fibers of the skin. This removal of the salt outside of the fibers causes an osmotic take-up of water into the hide fibers, and the skin will become rehydrated.

Parallel with rehydration is the removal of some of the globular proteins. These proteins, including the blood albumin and other soluble proteins, are dispersed by moderate concentrations of salt and are washed out of the skin at this time. The soluble proteins, if not removed from the skin prior to the tanning, will interfere with the tanning process. Upon soaking, the skins become noticeably softer and cleaner. After soaking they are often fleshed prior to liming.

Disinfectants

The removal of the curing salt and rehydration of the skin introduces the possibility of bacterial growth. The cured hide carries with it a large quantity of a wide variety of bacteria, and under the conditions of soaking they can become active. There are three areas of danger from such bacterial activity: the bacteria may be dangerous to man, infectious animal diseases may be present, and bacterial damage to the hides may occur.

Hides which are cured properly with clean salt, or are in a soft green salted condition and show no evidence of bacterial damage prior to soaking, generally have little necessity for the addition of disinfectants. Disinfectants of the type of chlorinated aromatic compounds are generally used. Disinfectants in the form of strong oxidizing agents are usually best avoided as they may interfere with subsequent unhairing processes.

Modern practice is to soak such a short time that bacterial damage is no longer a significant factor. Many of the earlier books on leather manufacture in their discussion of soaking include rather lengthy descriptions of possible bacterial damage and a discussion of bacteria in general. There are available a large number of different disinfectants, particularly of the synthetic type, so that there is very little difficulty in maintaining (for all practical purposes) relatively sterile conditions in the modern tannery. Concern for bacterial damage in soaking, therefore, is minimized.

Infectious diseases

Domestic hides are not considered dangerous, and ordinary rules of hygiene should protect the workers. Some imported hides may carry anthrax and should be handled with extreme care. Anthrax is a very serious and sometimes fatal disease which can be carried by hides and skins from some sections of the world. The disease may affect both animals and humans. It appears in humans as an open, ulcerated sore. If a worker develops open sores or boils and has been exposed to hides which could possibly contain anthrax, the physician treating the sores should be so notified. When properly diagnosed, anthrax can usually be controlled easily.

Some imported hides also may carry animal infections such as hoof and mouth disease. As a precaution Federal regulations require the use of specific disinfecting procedures.*

Temperature of soaking

In order to decrease the possibility of bacterial damage, soaking should be kept to a minimum, and most tanners feel it is advantageous to soak with as low water temperature as is available. Raising the temperature of the water, however, aids in the dispersion of the globular proteins and will accelerate the soaking process. Temperatures too high in soaking may result in looseness of the leather, veininess, and general coarseness of the fiber. The question of the temperature of the soak should be determined empirically and care should be taken to maintain consistency once the procedure has been established.

Detergents

Detergents may be used in soaking to aid penetration of the water and speed up the processes. There are three types of detergents: anionic, cationic, and nonionic. The nonionic detergents have proven to be the most effective. Both the cationic and the anionic types have a tendency to become fixed to the hide protein and carry over into subsequent operations.

Additives

Some tanners prefer to add a small quantity of sodium sulfide or sodium tetrasulfide to the soak prior to unhairing. This, in essence, is the beginning of the unhairing operation and may result in chemical action on the keratin (hair material) in the follicle and start the unhairing process. In the case of

*United States Bureau of Animal Industry, USDA.

sodium tetrasulfide, it has been shown that this assists in the removal of the soft globular proteins, which also has a bearing on removal of fine hairs.

Salt in the soak accomplishes two purposes: it aids in the removal of the globular proteins, and it aids in disinfecting the soak solution thereby decreasing the possible cause of bacterial degradation. If the soaking operation is conducted in a drum, the liquor ratio is such that the quantity of water is low; consequently, the salt washed out from the cured hide is generally sufficient to maintain salt concentrations at a proper level for effective removal of some of the soluble protein. In the case of dry hides or prolonged soaking systems, the addition of salt will permit the use of higher temperatures with decreased probability of bacterial damage.

Soaking of dry hides

Dry hides, or dry skins from the tropics, require considerably longer periods of soaking for complete rehydration. Since these skins may have been dried without the benefit of salt, there may be a partial gelatinization of the fibers, or an adhesion of one fiber to another. Mechanical action should be avoided in the early stages of soaking since the fibers are too stiff, and mechanical flexing will result in breaking of the fibers. Extensive damage to the skins can result if they are flexed too early before being soaked back. Quite often soaking periods as long as forty-eight hours are necessary to soak back dry hides and skins. It is quite common in these procedures to change water after twenty-four hours or more of the initial soak. The initial soak may contain a disinfectant or some salt, which will aid in the rehydration of the skins and allow prolonged soaking with reduced danger of bacterial attack.

The soaking procedure is closely related to the remainder of the beamhouse processes. Effective soaking greatly aids in removal of the hair and in subsequent unhairing processes, but the importance of proper soaking is very often overlooked by tanners in their attempt to improve the speed of product flow. Improperly soaked hides will take longer in the unhairing processes, and tanners are urged to be sure to soak a sufficiently long time and with effective methods in order to aid in the unhairing processes. Variations in soaking processes involve the use of additives, changes in temperature, degree of agitation, and length of time of soaking.

UNHAIRING

The unhairing of hides and skins may be accomplished by a number of different means involving widely different principles of operation. All means are, however, related to the chemistry of hair and the soft keratin proteins in particular. In the earlier consideration of the structure of skin and the growth of hair it was pointed out that hair grows in the hair follicle, and at this point there is a transition between the liquid protein building blocks which feed the hair cells in the follicle and the formation of fibrous structure making up the shaft of the hair. This area of high metabolic activity is also one in which the keratin is most reactive chemically. Therefore, unhairing procedures are most effective in the hair follicle. Removal of the hair at the root can be done by a number of different chemical methods.

Liming

Lime (CaO•H$_2$O) dissolves in water to the extent of .1 parts per 100 parts of water. This limited solubility results in a solution of pH near 12.5 (more or less, depending upon temperature, water condition, and lime quality). This buffered alkaline medium is easily obtained and has been used as an unhairing system since ancient times.

Hair loosening. The action of lime on proteins results in hydrolysis of the protein, a gradual breakdown of the the the protein structure, and solubilization of the proteins as smaller and smaller molecules. An inspection of the amino acid content of the three major types of protein materials found in hide indicates the differences in the chemical action of lime. In the case of globular proteins, the high degree of solubility is due to the prevalence of free acid and amino groups in the skin. In the case of collagen, fewer acid and amino groups are available, and the collagen will only be dissolved by prolonged action of strong acids and alkalis. Keratin is also made up of amino acids containing both acid and basic groups. Keratin is distinguished from collagen in that it has relatively large quantities of the sulfur-bearing amino acids, e.g., cystine and methionine. Hair is more resistant to acid than hide collagen.

The keratin can be changed by the addition of reducing agents; the sulfur-bearing proteins are released to form SH groups on the protein. The SH groups can then act as acid groups in relation to alkali in the solution. Reduced keratin, therefore, has greater susceptibility to hydrolysis, and in this condition is more soluble in alkali than collagen. The addition of both an alkali and a reducing agent results in a preferential breaking down and a solubilization of hair keratin without damaging chemical action on the leather-making collagen. This is the basis for most lime unhairing systems.

On the other hand, elastin, another major component of hide protein, has very few active acid or alkaline groups and has very little tendency to be solubilized by the liming operation. The removal of elastin, therefore, is accomplished later in the beamhouse processes during the bating operation.

Alkali Immunization. Hair removal by the combination of reducing agents and lime based on a hydrolysis of the crystine link can be greatly impaired by the application of excessive alkali prior to the application of the sulfide. The theoretical mechanism of this particular reaction in the hide protein has been argued by leather technologists without a universally accepted explanation. In a practical case, if an excess of alkali or lime comes in contact with the hair in the follicle prior to the reaction of reducing agents, the action of the reducing agents is greatly limited. This phenomenon has been known to the industry as alkali immunization.

Other effects of the liming operation. The function of the beamhouse process is not limited to the removal of the hair. The action of the chemicals on the hide changes the chemical and physical nature of all components of the skin. The removal of the globular proteins, which was stated in the soaking operation, is carried further toward completion during the liming operation by the action of the alkali. The action of the alkali and reducing agents on the elastin is somewhat limited, but it may contribute significantly to the ease of the removal of the elastin by the proteolytic enzymes using in bating. The presence of the alkali in the skin during the liming aids in the saponification of the natural fats and other fat-type materials.

Effects on natural fats. The natural fats of the hide fall into two general classifications, as mentioned earlier. First, the physiological fats necessary to the maintenance of body temperature and lubrication of the hair contain rather complicated lipid structures such as the phospholipids, lecithin, cephalin, and other lower melting fats. The second classification of fats are the triglycerides of the fatty acids which have, in general, higher melting points and are more difficult to solubilize. During the liming operation, lecithin and cephalin will be easily removed, but the triglycerides may persist longer due to the relative insolubility of the calcium soaps. Removal of fats during liming or during alkaline treatment in the beamhouse processes is accentuated by increase in temperature and alkalinity through the addition of sodium sulfide or other strong alkaline materials. The use of high temperatures and high alkalinities in liming, however, may result in rather drastic effects on the hide which far offset any benefit which may be obtained in grease removal.

Swelling effects. In the previous outline of the biochemistry of hides, skins, and skin proteins it was pointed out how the swelling of hide protein or hide collagen is affected by the changes in pH and salt concentration in a skin. Normal liming operation takes place between pH 12.0 and 13. When the hide is introduced into the lime, the grain and flesh surfaces of the skin are subject to solutions at pH approximately 12.5, while the center of the skin is still near neutral. A differential swelling occurs between the grain, flesh, and center of the skin in accordance with the variation in pH existing at that time. As liming proceeds, the pH differential becomes less and less and the skins become uniformly swollen. In general, hair removal may take place at a much faster rate than equilibration of the pH differential, and as a consequence, in some liming processes equilibration of the swelling may not take place. The action of lime or other alkali on hide protein will also result in the disruption of some of the hydrogen bonding between the adjacent protein chains. This will make available more acid and alkaline groups on the skin as evidenced by the change in the pH titration curve of the hide protein. Practical tanners refer to this as the ''opening up'' of the hide during liming. This phenomenon is both physical and chemical. The longer the liming, the more opening or fiber bundle splitting may take place in the skin. Parallel with this will be an increase in the chemical activity of the skin. The action of this prolonged swelling has a definite practical effect on the hide with respect to its leather-making properties.

In accordance with the base-binding capacity of the protein, an equilibrium will be established between the protein fiber and the solution immediately surrounding it. When a hide is placed in a solution of strong alkali, some of the alkali will be absorbed by the skin, resulting in an increase of the pH of the skin and a decrease in pH of the solution. The proteins at the surface of the skin absorb the alkali, and in effect, the concentration of the alkali acting on the grain surface of the skin may be several times stronger than the alkali acting on the center layers of the skin. A differential swelling will result in internal stresses and in some cases may result in rupture of some of the fiber structure of the skin. This will permanently affect the nature of the grain of the leather.

Extensive studies of the swelling of hide fiber during liming operations have been made by histological means. Also, the swelling of the hide proteins has been observed by chemical analysis using the stratigraphic technique. Both these methods have their shortcomings: interpretation of the histological method is very difficult by anyone but an expert in the field, and in the stratigraphic technique the amount of data necessary for a single observation represents a tremendous amount of work. Interpretation of stratigraphic data, too, is complex.

The hydration (or swelling) of a protein results in a much greater increase in the diameter of the fiber than in its length. The orientation of the fibers in the skin is not uniform from grain to flesh, but there is a distinct change in the angle of weave. Differential swelling effects may be further accentuated and may cause a physical distortion in the shape of the fiber. The fibers on the flesh side, in general, run parallel to the surface of the skin. The fibers on the grain side, on the other hand, run more or less perpendicular to the surface of the skin, or more likely, parallel to the direction of the hair folli- cle. Since the fibers swell more in diameter than they do in length, there will be a greater expansion of the surface area of the skin on the grain side than on the flesh side, which will result in a curling or rolling of the side. This oc- curs generally along the backbone, and the skins may roll up quite tightly into long, ropy, sausage-like bundles. During this rolling action, improper liming may occur as a result of lack of penetration of additional alkali, and damage may be done to the skins along the backbone area. In order to avoid this, the amount of mechanical action used in the initial stages is limited un- til after the skins have obtained a more uniform swelling. Once the swelling has equilibrated, the formation of "rollers" is greatly decreased. If cattle hides are not sided prior to soaking or liming, "rollers" will not form. This is the accepted modern practice.

Action of the liming on the hair. The hair structure consists of the hair bulb in which the soluble proteins of the hair are picked up into the hair root; the shaft of the hair, of relatively constant diameter, in which the hair has formed an inert, crosslinked keratin protein; the tip, at which the diameter of the hair ta- pers down to a vanishing point. The action of chemicals on the hair is quite

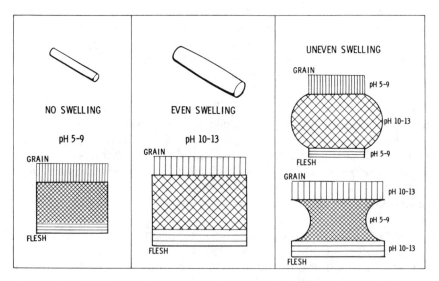

Figure 2 Effects of pH on swelling. An individual fiber will swell more in diameter than in length. At pH above 10 swelling will occur, but between pH 5 and 9 there is very little swelling. If the pH through the skin is uneven, distortion will result and stresses develop within the skin.

different in the three areas. In the follicle the keratins are soft, i.e., they have not yet formed complete cystine crosslinkages and therefore have not attained an inert structure. In the hair shaft the hair is most inert, and due to the scale on the surface is little subject to chemical activity. On the hair tip the small diameter exposes a larger surface area for unit weight, and a greater portion of the hair is exposed to chemical action. The behavior of the hair toward alkali and reducing agents is quite different, depending upon the strength of the reducing agent and the specific portion of the hair involved. If a very mild reducing system is used, for example very low quantities of sodium sulfide or sulfhydrate, the effect of the alkali would be dominant. Hydrolysis of the soft keratins in the hair follicle would allow removal of the hair from the hair root without significant chemical action on the shaft or tip. The hair with good strength and in a usable form could then be recovered.

In the case of a very sharp liming system, i.e., large quantities of strong reducing agents such as sodium sulfide, the action of the strong reducing agent can cause a breakdown of the sulfur-sulfur linkages; this action takes place primarily at the surface of the hair. Since greater surface area is exposed on the hair near the tip, the hair is destroyed progressively from the tip on down toward the root. In such liming systems, where sufficient alkali and strong reducing agents are used, the action is very rapid and destruction of the hair takes place in a matter of minutes after the beginning of the liming operation. If the action is not quite sharp enough, destruction of the tip of the hair and a distortion of the fiber of the hair (curling) may take place. The partially destroyed hair then becomes matted and may be very difficult to remove in subsequent liming.

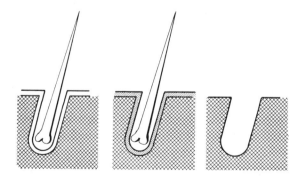

Figure 3 Action of sharpening agents during a hair-saving system in a mild liming system. The epidermis and soft keratins in the hair follicle are slowly dissolved and the hair is removed by the root, leaving a clean follicle.

Windus and Showell have reported the explanation of mechanism of unhairing as a nucleophilic displacement. The general basis of the nucleophilic reaction is the exchange of organic components by a donation of electrons by a nucleophile. The general rule of the reaction is as follows:

$$R-S-S-R + 2Na_2S \longrightarrow 2R-S \cdot Na + Na_2S_2$$

When interpreting this as an unhairing action and the action of the sulfides, the following reaction can be considered:

$$R-S-S-R + S^= \longrightarrow R-S-S^- + RS^-$$

$$RS-S^- + S^= \longrightarrow R-S^- + {}^-S-S^-$$

The difference appears to be slight. However, the nucleophilic approach can be used for the explanation of the unhairing action of dimethylamine by the following reaction:

$$R-S-S-R + (CH_3)_2-\ddot{N}H \longrightarrow RS:N(CH_3)_2 + RSH \cdot NH(CH_3)_2$$

The unhairing activity of specific materials can be listed in decreasing order, and this is consistent with their ability to act as nucleophiles. The activity of a material to act as a nucleophile depends upon its pH. Using this new theory as a basis should point to the development of new unhairing agents.

$$\genfrac{}{}{0pt}{}{S^=}{RS^-} > CN^- > (CH_3)_2\ddot{N}H \quad > S_2O_4 > OH^- > \genfrac{}{}{0pt}{}{SO_3{}^=}{S_2O_3{}^=}$$

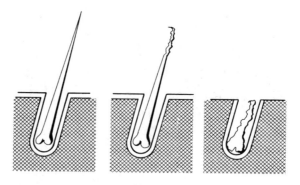

Figure 4 Action of sharpening agents in a very strong liming system. The chemical action is strong enough to destroy the hair. The tip of the hair, having the greatest exposed surface, is destroyed first with the rest following. Softened hair shafts remain in the hair follicle.

Materials used in lime unhairing. (a) Sulfide and sulfhydrate: Sodium sulfide, upon hydrolysis, forms sodium sulfhydrate and caustic soda at the normal pH of liming. For each mole of sodium sulfide introduced, one mole of sodium sulfhydrate and one mole of caustic soda result. The use of sodium sulfide, therefore, to give increased sharpening also results in increased alkalinity, which may have undesirable effects with regard to the swelling of the hide. Sodium sulfide is considered a very sharp liming agent and rather harsh in its action, both on hair and hide protein. For a milder action, sodium sulfhydrate is used. Sodium sulfhydrate was developed specifically for use in the leather industry. It has the advantage of allowing the introduction of increased reducing action without raising the pH. A balance between the use of sodium sulfide and sodium sulfhydrate will give the tanner a wide range of alkalinities to be used in the liming operation.

Sodium sulfide and sodium sulfhydrate are the most dangerous materials used in the tanning industry. Upon acidification, solutions containing sulfides, even in trace amounts, will release hydrogen sulfide gas into the atmosphere. The gas is heavier than air and may accumulate in sewers or wherever solutions from acid processes and sulfide unhairing systems come together. Almost every year, the industry has fatal accidents. Hydrogen sulfide, at very very low levels, has a noxious odor, which serves as a warning. At high enough levels to be fatal, however, hydrogen sulfide paralyzes the nerves, and fatality may result without warning.

(b) Arsenic sulfide: Arsenic sulfide has been used in the liming operation and is considered to be a relatively mild reducing agent. This material will allow the removal of the hair with little raising of the pH. Although arsenic sulfide was rather popular some years ago, its popularity has been greatly decreased because of the slowness of its action and the possible toxic effects resulting from its use.

(c) Sodium cyanide: Sodium cyanide also has been used as a mild sharpening agent. Hair destruction is generally not encountered with cyanide, and it is an effective unhairing agent. Because of its toxicity, cyanide is generally applied in a two-bath process in which the first bath is lime and cyanide and the second is plain lime. The use of cyanide has been avoided by most tanners because of its high toxicity. It must be pointed out however, that hydrogen sulfide gas has a lethal toxic level approximately equal to that of hydrogen cyanide.

(d) Sodium hydrosulfite: This is another compound which has mild reducing properties and which can be used in blends with sodium sulfide, sulfhydrate, or other sharpening agents to produce excellent unhairing.

(e) Dimethylamine sulfate: Early observations of practical tanners indicated that old limes, i.e., limes which had been re-used, were mellow and resulted in good unhairing with very little harsh swelling action. In a search for the chemical explanation of this phenomenon it was found that old, mellow limes had available various amines which were attributed to the breakdown of the hide proteins. In a comprehensive study of the effects of amines in unhairing it was evident that dimethylamine was a most effective unhairing agent. This has led to the use of dimethylamine sulfate as an unhairing agent and to the production of this chemical specifically for use in the leather industry. It is a mild sharpening agent and will aid in unhairing in a lime system. It also results in a lowering of the pH during the liming operation.

Many unhairing systems in practical use are based on a balance between sodium sulfide, sulfhydrate, dimethylamine sulfate, and sodium hydro-sulfite to produce rapid, efficient unhairing systems with controlled swelling.

The BASF Corporation has developed a new system which they call the Molescal System. This is a proprietary product which has the property of being an active sharpening agent but also, it is easily oxidizable and very little sulfide will remain in the system. The Molescal System can be used with or without the addition of a small quantity of sulfide to aid in the unhairing and can be used for the manufacture of both hair save and hair burn leathers. This greatly reduces the sulfide effluent which is attractive to tanners particularly with effluent problems. The leather quality is equivalent to that from a lime sulfide hairing system.

Practical unhairing systems—painting system (wool pulling). In the unhairing of small skins, particularly those with large quantities of hair such as sheepskins and many goatskins, the painting system is used. The system, as described here, is typical of the sheepskin industry. The sheepskins, either fresh, salted, or brined, are washed and sometimes centrifuged prior to painting. The painting is conducted with a solution of lime and sodium sulfide, sodium sulfide alone, or sodium sulfide plus caustic soda. The strong reducing agent is painted on the flesh side of the skin and skins are piled in such a manner that the sulfide paint is sealed in. Care should be taken so that the wool does not come in direct contact with the strong sulfide painting solution. The skins are piled in a cool place, or hung over sticks, or folded along the backbone, generally overnight. Next day the skins are ready for pulling. In pulling the wool, the skin is laid over a beam with the wool side up, and the wool puller removes the wool by hand. Since the length of the fiber of the wool is different for the different areas of the skin, and since there may be burrs, paint, or other defects in the wool, the puller removes the wool from the various areas of the skin and places it in different quality classifications. The unhaired skin, or slat, is then placed in an additional liming solution without sharpeners, usually overnight, and this is followed by bating and pickling.

The painting process may also be used with goatskins or calf. The object here is not to recover the hair, but rather to get a thorough removal of the hair without soaking in a lime slurry. In light skins the hair-to-skin ratio is rather high and with an immersion sysem large quantities of sharpening agents may be necessary. In the painting system, however, the concentrated sulfide-lime material penetrates through the skin from the flesh side and may destroy the hair in the root. The painting-liming system would not be effective on heavy cattle hides due to their thickness. The painting system is time-consuming and laborious. Its use is justified only where the value of the hair is very high, or where it has particular quality effects on the leather being produced.

Calfskin liming. The value of calfskins is in their very fine grain and excellent surface characteristics. Liming, therefore, should be very mild. Two general liming systems are used for calfskin. The first is the painting system outlined above in which the strong alkali and strong reducing agents are in contact with the skin for a limited period of time. The other is a milder, slower system. Three or four days' liming for calfskin is quite common, and many processes are used in which the liming may continue much

Figure 5 Sheepskins are normally unhaired with a painting system and later the wool is pulled by hand. The operator places the different grades of wool in different containers. This hand operation is standard practice. *(Courtesy Morris Fishman.)*

longer. Calfskins are often given an initial liming of mild sodium sulfide and sulfhydrate for about two days, followed by unhairing, and then by a second liming without any sharpeners for another two days. The resulting skin is very clean, free of fine hair, and in excellent condition for subsequent tanning operations.

Upper leather

Hair-saving. The hair-saving system involves the introduction of sharpening agents with the lime in a balanced system, with careful attention to temperature and concentrations. By proper choice of materials, unhairing systems of one or two days' operation will result in clean unhairing for shoe upper leather. Removal of the hair takes place by hydrolysis of the soft proteins in the hair follicle. This method was very common in the United States and results in the valuable by-product of the hair itself. The system is generally conducted in paddles.

Hair destruction (burning). The main advantage of the hair-destruction system, as practiced in the United States and Europe, is the speed of opera-

Figure 6 Beamhouse paddles. These wooden vats are used for soaking and liming hides and are equipped with hot water, cold water, and steamlines. The temperature can be adjusted by the addition of steam in a protected box. These paddles have a capacity of 10,000—12,000 pounds of cattle hides, or about 200 hides in 5,000 gallons of liquid. *(Courtesy Johnson and Carlson).*

tion and lower costs. Under a hair save system, heavy limited hides are unattractive. It is very difficult to find people to do this type of work.

In the hair-destruction system the quantities of sodium sulfide and sodium sulfhydrate used are sufficient to cause the destruction of the hair. In this process the skins are introduced into a drum and sodium sulfide, sulfhydrate, and lime are added. Very rapid destruction of the hair results, and within fifteen minutes the hair loses most of its fiber structure. The hair-destruction system may result in uneven swelling and in the formation of scud (surface dirt) on the hides. Properly conducted though, the system results in a very clean surface and a tight grain. The hair-destruction system is quite common in leather production for all types of leather.

By using a hair-burn system, in either a hide processer or large drums, the liquor ratio is very low. The amount of water to the hide is kept quite low, and as a result, there is a strong rubbing action of the hides over one another during the unhairing process. The rubbing action removes the hair and the surface dirt very effectively. It is common practice to do the hair burn with a very strong sulfide solution; that is about 30 pounds of sodium

Figure 7 In a hair-saving system the hair removal is accomplished with an unhairing machine. Through proper liming the hair should be recovered in usable form with clean hair removal from the skins. *(Courtesy Granite State Leather Co.)*

sulfide or sodium sulfhydrate, or more, per 1000 pounds of hide. After running in this solution with perhaps 1 or 2% hydrated lime, for less than two or three hours, the hides are washed by flow-through water to remove the sulfides, and additional lime is added. Added lime will decrease the very harsh alkaline conditions that were employed during the hair burning phase. The pulped hair and hair burn solutions, as they are dumped, are the foremost problem in effluent disposal in tanneries.

Hide processors have become the standard method of unhairing in the hair-burn system. The hair burn is used for cattle hides for labor efficiency and for better effluent control. The value of the hair is not sufficient to cover the costs in temperate climates even though the hair of the cattlehides may be thick and heavy. In the tropics the cattle hair is not of good enough quality to save. If goatskins are processed the hair is much more valuable so hair save systems in paddles are more common.

Figure 8 Hide processing machines. These "cement mixers" are used in many modern tanneries for soaking, liming and unhairing. In some tanneries the hides are placed in the processors and are not removed until after chrome tanning. The system has the advantages of high speed, ease of control, automation, and lower labor costs. The processors can be emptied by reversing the direction of rotation. Loads in excess of 20,000 lbs of hides are common for this type of equipment. *(Courtesy of S. B. Foot Tanning Company)*

Non-lime unhairing systems

The use of lime as an unhairing agent is so common in the leather industry that the terms "beamhouse," "liming," and "unhairing" are practically synonymous in their meaning. There are, however, several inherent disadvantages to the liming system that have led to the search for other practical methods of unhairing. This intensive research has yielded several practical methods of unhairing and has pointed the way to other systems which will probably gain in popularity in the future. One of the main reasons for such searches is the elimination of lime, and possibly sulfide, in order to alleviate the problem of effluent disposal and stream pollution. The systems which have been proposed in research work have been very broadly based, and only a few have attained any degree of commercial acceptance. We cannot, of course, discuss all these systems, so we will confine our remarks to those few that have been given the greatest practical attention in recent years.

Chemical unhairing systems. We can divide the chemical unhairing systems into two classifications: those based on alkaline reducing systems and those based on oxidation.

In the United States, the *alkaline* reducing systems have been given great attention. Studies have been made of the unhairing properties of ammonium hydroxide and buffered alkaline systems from a research point of view, and few practical systems have been employed on an experimental basis. Although these systems have given reasonably good unhairing, the overall advantages of quality improvement of the leather, high speed operation, economy of chemicals, and simplicity have not surpassed the lime unhairing systems on all counts, and therefore, general commercial acceptance has not followed.

The most intensive work in recent years on a commercial scale has been in the pursuit of the dimethylamine sodium hydroxide system, as proposed by the Rohm & Haas Chemical Company, particularly Dr. Somerville and his co-workers. Dimethylamine has long been known as an effective unhairing agent when used in lime systems. Dimethylamine caustic soda sodium sulfate can, in proper concentration, effect unhairing with a hair-saving system. The proper maintenance of a balance of caustic soda in the system is of great importance, and this is done by a control system involving the titration of the sodium hydroxide in the solution. As a result of this work, a 24-hour hair-saving procedure employing dimethylamine sulfate (DMAS) and caustic soda has been used on calfskins and cattle hides to make both lightweight leathers and heavy sole leathers. One of the main drawbacks of the system is that the stock is quite slippery due to the caustic soda, which makes it much more difficult to handle.

In the interest of economy, and also to decrease the strong action of the caustic soda, the system has been modified to incorporate DMAS, lime, soda ash, and sodium sulfhydrate. By using a balance of these materials, the harshness of the caustic soda is overcome by a buffering action attributed to the lime, and additional unhairing results are reported in both fresh and cured hides. Further development work has been done so that either hair-destruction or hair-saving systems can be employed based on the balance of these components.

Oxidative system. The unhairing systems discussed so far have been based on the alkaline hydrolysis of the keratins after the action of reducing agents to break the disulfide linkage, or the action of dimethylamine on the sulfide bonds of cystine in the hair protein. Because of the objection to the sulfide ion in the effluent from the tannery, and also due to the tendency of the sulfides to form colored insoluble compounds, practical processes which can eliminate the sulfide would be most welcome to the industry. A further problem in tannery effluent is the discharge of the lime. As a possible means of solving both of these problems, a system has been developed in Germany by Dr. Rosenbusch of Farbwerke Hoechst, in which the unhairing is done by a strong *oxidizing* medium under acid conditions.

The active ingredient in oxidative unhairing is chlorine dioxide which reacts with the sulfur-sulfur bonds of keratin to produce a keratin sulfonic acid and free chlorine. The keratin substance of the epidermis is then converted into water-soluble materials, and in the weakly acid medium the hair and the epidermis are removed by the mechanical action of the drum itself. The chlorine liberated from the reaction, is absorbed by the collagen. It is claimed that a large quantity of the natural fat is saponified and a loosening of the fiber structure of the hide is achieved so that the unhaired hides can be pickled and ready for tanning in one operation, in about twenty-four hours. The medium used is glycolic acid, which maintains a buffered pH between 3 and 3.5. In order to maintain the proper pH value, sulfuric acid is added. The temperature is below 40° C in order to prevent any hydrolytic decomposition of the collagen, and slowly rotating drums (4-6 rpm.) are used to prevent the generation of too much heat. The strong oxidizing action of the chlorine dioxide and chlorine results in the bleaching of the hair, and there is no dark scud left on the hide. After completion of the unhairing the excess oxidizing agent is decomposed with thiosulfate in order to prevent the strong oxidizing agent from being carried over into the chrome tanning bath. The strongly oxidizing chlorine dioxide would oxidize the trivalent chromium tanning agents to dichromate. The addition of the thiosulfate to the unhaired stock will result in the formation of some sulfur within the skin, which is desired in some types of leather manufacture. It is also claimed that the absorption of the chlorine by the hide protein has the advantage of giving a fine grain and making a somewhat firm leather. The excess oxidizing agent must be reduced with thiosulfate prior to the use of vegetable tannins, in order to avoid excessively rapid fixation of vegetable tanning agents. Chrome tanning can be conducted in the usual manner.

The oxidative system has drawbacks in the high cost of materials and the necessity of shifting production from an alkaline system to an acid system, which would mean a complete readjustment of the subsequent processes of tanning, coloring, and fatliquoring. Considering the problem of tannery effluent, however, the prospect of an acid oxidative unhairing system is intriguing. It will be most interesting to watch the progress of this development.

TNO system. Van Vlimmeren and his co-workers at the Institute for Leather Shoe Research (TNO) have developed a very interesting unhairing system which involves a somewhat different mechanism. Van Vlimmeren was interested in developing an unhairing system which would allow coagulation of the soluble protein removed during the unhairing. This would alleviate serious effluent problems.

Van Vlimmeren worked with sulfide and caustic soda in the unhairing system, followed by liming to give the proper swelling and plumping of the skins and hides. He found that the drastic action of the caustic soda resulted in poor quality leather due to the wrinkling of the grain, and the leather also tanned out quite hard without the action of lime. It was reasoned that the presence of calcium ion in the skin had a buffering action and also a specific

TNO soaking—unhairing procedure.
(percentages based on salted weight; chemicals on 100% basis)

Soaking:
  100% water 20°C
  1.0% calcium chloride (as $CaCl^2$)
  0.4% magnesium oxide
  Drum 1 h. at 2 r/p/m
  Drum 5 min/h. for 15 h.

Green—fleshing:
Unhairing:
  30% water 20°C
  1.8% sodium sulfide
  0.8% sodium hydroxide (dissolved in 8% water)
  Drum ½ h. rest ½ h. etc. for 3½ h.
  Add 60% water 20°C
  Drum 5 min/h. for 16 h.

Washing:
  100% 20°C, 3 times

ion action on the hide protein which was beneficial for quality leather. Van Vlimmeren was interested in decreasing the total quantity of calcium in the solution and the quantity of lime for effluent control purposes.

To solve this problem, the calcium was added at a relatively low pH and allowed to penetrate into the hide during the soaking period. After the soak, which was a relatively high pH, around 10, the solution could then be drained off and the unhairing proceeded with the addition of sodium sulfide and sodium hydroxide. All of this was done in relatively low water content. The TNO experiments indicated that the presence of magnesium oxide in the calcium chloride soak gave the desired pH. The specific ion effect of the magnesium gave desirable effects. TNO reported that the physical properties of the area and the yield were not significantly altered from conventional processes. The cost of the operation was similar to those in the conventional processes. The author has run this system in several client's tanneries with good results. The significance of this process on effluent treatment will be discussed in the chapter on tannery effluent.

Enzymatic unhairing. Enzymatic unhairing may be considered to be one of the newest and yet one of the oldest practical methods of unhairing. Enzymes are very complicated and specific catalysts of nature upon which the life processes are based. They may be separated from the living organism, or they may be removed from the organism after it has died. Commercial enzymes are from plant, animal, or microbiological sources. In our discussion of the decomposition of hides by rotting (see Chapter 3), we stated that the areas of high metabolic activity, those containing proteins of low molecular weights, were the areas first attacked by microorganisms. Enzymes excreted from microorganisms or applied under controlled conditions by the tanner may selectively bring about the hydrolysis of the proteins in the hair follicle. This will result in a controlled hair slip and removal of the hair in good condition. Enzymatic unhairing was accomplished by means of controlled bacterial action in primitive tanning systems. The American Indians soaked hides in water and allowed them to start to rot, at which time the grain was peeled off at the junction between the grain layer and the corium at the base of the hair root. This resulted in the formation of a suede-type buckskin which was common wearing apparel for the Plains Indians.

The sweating system of hair removal was applied by dipping the skins in water and hanging them in a moist room for a period of time until the hair slip took place by natural wild bacterial growth. The system resulted in a clean unhairing, but more often than not excessive damage to the hide collagen resulted and pits and holes were found in the leather.

As a result of modern chemical technology, it is possible to obtain enzymes of very specific chemical activity. A wide variety of enzymes can be used for unhairing. In some of his early research work on the subject Burton found that amylase, a starch-splitting enzyme, could be used for enzymatic unhairing, and as a result, much of the research work was concentrated in this direction. Cordon, of the United States Department of Agriculture, investigated a number of enzymes and found that many of them could be used for enzymatic unhairing and were of both amylytic and proteolytic activity. Nickerson, working at Rutgers University, developed a keratinase, an enzyme specific for its action on keratin, and this has been applied commercially in an extensive investigation of enzymatic unhairing. In investigating enzymatic unhairing, practical systems were developed in the early 1930's based on Aspergillus orizae. This was particulary used for the enzymatic unhairing of goatskins, both in the United States and in Europe.

Additional enzymatic unhairing systems have been developed which are compatible with the high alkaline pH conditions and lime which result in a hair removal system obtained with the enzyme and a relatively high pH. The manufacturers claim that the leather quality is much better than in other enzymatic systems in that tight grain and full leather can be made. The use of enzymatic unhairing whether lime condition or a neutral state usually results in a general loosening of the grain of the leather which makes the process quite undesirable.

The author has investigated enzymatic unhairing extensively in an attempt to develop a practical, economic means of hair removal. Enzymatic unhairing has the advantage of being able to remove the hair in good condition and high strength, although it may decrease the felting properties of the hair. The skins are kept at non-swelling pH value so that very tight leather can be produced. There is no objectionable effluent problem resulting from such unhairing. Temperatures employed are usually at 90° F or above, although in some of the more recent developments lower temperatures have been employed. There is no swelling of the collagen during enzymatic unhairing, and this has been one of the main drawbacks to its practical acceptance by the tanning fraternity. Enzymatically unhaired hides, when tanned by conventional procedures, will result in fine grained, tight leather with a tendency to be tinny, hard, and thin. Extensive modifications of the tanning, coloring, and fatliquoring procedures are necessary to compensate for the drawbacks. One of the simplest ways of correcting the disadvantages of enzymatic unhairing is to introduce an alkaline swelling bath, of either lime, caustic soda, or soda ash prior to the pickling and tanning operations.

The cost of enzymatic unhairing has been a drawback in some instances. This author, in his work on enzymatic unhairing, took the premise that cost was the most important factor, and consequently set about to develop a low-cost unhairing system prior to investigating the modifications that were necessary to make it compatible with existing tanning processes. It was found that by using a mixture of sodium sulfide, sodium bisulfite, and ammonium hydroxide, a buffered medium at a pH of about 9.5-10 was obtained in the presence of strong reducing agents. The enzyme papain introduced into this system exhibited very strong unhairing action, even at temperatures as low as 80° F. Excellent unhairing results were obtained, but it was necessary, in order to overcome the typical enzymatic characteristic of the leather, to modify the tanning processes by intermediate alkaline treatments with caustic soda or lime, and also to modify subsequent tanning processes.

The general characteristics of leather from enzymatic unhairing are loss in thickness and increase in area, particularly in length from head to tail. Enzymatically unhaired leather also has a very fine grain and a tendency toward crackiness.

LOOKING AHEAD

The beamhouse processes are a combination of several phenomena: soaking, removal of the globular proteins, removal of the hair, and opening of the fibers for desired leather character. All these processes should be considered in their interrelated positions. The effects of soaking are additive on liming, the effects of liming are additive on the enzymatic activity of bating. The practical tanner can greatly improve his processes by thinking of these systems as an integral unit rather than as isolated cases. The addition of a small quantity of proteolytic enzyme into the soak could greatly shorten the unhairing time, particularly in a hair-saving system. The quantity of lime used in unhairing can be balanced with other chemical agents, using modern chemical controls to improve the problem of effluent from the tannery. It is through a combination of chemical and enzymatic systems that the beamhouse of the future will operate.

In the 1990's, the effluent problem, particularly as regards sulfide, continues to dominate. To date, however, the approach of oxidation unhairing, the TNO system, and other processing changes have not achieved significant commercial acceptance.

The acceptance of a radical change in beamhouse methods is a slow process for the practical tanner. Each step in the production is dependent, both physically and chemically, on the preceding steps. Any change in the beamhouse methods will require changes in the subsequent processes. These changes will result in further changes so that the entire production system must be altered. Another problem is that beamhouse changes will affect all of the production of the tannery, thus putting the entire production at risk. If a new finish or a new retannage is contemplated only a limited quantity of leather is at risk. Such changes are welcomed in the search for quality, reduced cost, or customer satisfaction. Major changes in the beamhouse methods are made only in a conservative way.

REFERENCES

Goetz, A. W., "The Chemistry and Technology of Leather" (O'Flaherty, Roddy, and Lollar, ed.), Vol. 1, New York, Reinhold, 1956.

Gustavson, K. H., "The Chemistry and Reactivity of Collagen," New York, Academic Press, 1956.

Gustavson, K. H., "The Chemistry of the Tanning Process," New York, Academic Press, 1956.

J. Am. Leather Chemists' Assoc., Supplement No. 2, Round Table Discussion (1952).

J. Am. Leather Chemists' Assoc., Supplement No. 3, Leather Making (1952).

"Leather Chemicals," Philadelphia, Rohm & Haas Company, 1955.

Lollar, R. M., "The Chemistry and Technology of Leather" (O'Flaherty, Roddy, and Lollar, ed.), Vol. 1, New York, Reinhold, 1956.

McLaughlin, G. D., and Theis, E. R., "The Chemistry of Leather Manufacture," ACS Monograph 101, New York, Reinhold, 1945.

Merrill, H. B., "The Chemistry and Technology of Leather" (O'Flaherty, Roddy, and Lollar, ed.), Vol. 1, New York, Reinhold, 1956.

Morris, W., "The Chemistry and Technology of Leather" (O'Flaherty, Roddy, and Lollar, ed.), Vol. 1, New York, Reinhold, 1956.

Rodgers, A., "Practical Tanning," New York, Reinhold, 1956.

Stiasny, E., "Gerberichemie," Steinkopff, 1931.

Turley, H. G., and Windus, W., "Stiasny Festschrift," Darmstadt, Eduard Roether, 1937.

DELIMING, BATING,
AND PICKLING

7

The purpose of deliming, bating, and pickling is to prepare the unhaired hides, chemically and physically, for tanning. Limed hide has a pH of about 12, the fibers are swollen and distended, and the hide contains undesirable chemicals and hide degradation products from liming. Deliming, bating, and pickling are further steps in the purification of the hide collagen network. Deliming is the removal of alkali and the adjustment of pH for bating. Bating is an enzymatic action for the removal of unwanted hide components. Pickling is the adjustment of the pH of the skin to the level desired for tanning or hide preservation. The three processes are usually conducted as sequential steps in a single batch operation.

DELIMING

Limed hide is a three dimensional network of protein fibers. These fibers have absorbed alkali in the form of lime and other alkaline material. The skin is a blue-white, swollen, rubberlike, semi-translucent mass. In deliming, the three dimensional character of the skin is of great importance, and different conditions of pH and swelling may exist from one part of the skin to another. For effective bating and cleaning of the skin a slightly alkaline pH is desired.

The deliming process, therefore, must be one that will solubilize the absorbed calcium hydroxide and bring the skin to the desired pH. The solubility of calcium hydroxide is approximately 0.1%, decreasing with increasing temperature. The hide protein has absorbed calcium hydroxide which must be dissolved as a soluble calcium salt during the deliming operation. Additional acid is necessary to bring the pH down to the desired level because of the acid-base binding capacity of the protein. The hide-lime system, therefore, is buffered both by the lime and the hide protein. If, by calculation or experiment, we were to determine the acid necessary, and were to add hydrochloric acid to the limed skin, the pH at the surface would be very low while the pH inside the skin would be high during the initial reaction. Since acid swelling would occur at the surface layers and alkaline swelling in the center, there would be sections in the skin below the surfaces in which there would be very little swelling near the isoelectric point. The resulting mechanical distortion would probably rupture the fibers to some extent, and the fine grain quality of the leather would be damaged. The deliming process must take place very gradually by closely controlling the pH to the range of little swelling at the isoelectric point. This is done by the use of a proper balance of acids and buffering salts in which the lime may be solubilized and easily removed.

Calcium has good solubility in ammonium sulfate or ammonium chloride. The natural buffering pH of ammonium sulfate is about 5. In the presence of the calcium hydroxide and ammonium salts the solution pH is buffered between 7 and 8; good solubility of the lime results. The lime is gradually removed from the skin by diffusion, and the swelling effects on the hide during this operation are kept at a minimum. This is also the pH range of maximum enzymatic action during bating.

There is now concern about the possible toxic effects of the ammonium salts in the environment. The concentrations of ammonium discharged by the tanning industry are in excess of what is acceptable under some regulations. The search for an ammonium-free deliming system has led to the use of magnesium salts as deliming agents.

Magnesium salts are very soluble. Magnesium sulfate has a solubility of 27–68g/100ml water, depending on the temperature. Magnesium hydroxide is very slightly soluble. (less than .01 g in 100ml cold water). Magnesium hydroxide, in a slurry, will develop a pH of 5–7 depending on other materials in the solution. When a limed hide is placed in water containing magnesium sulfate, the alkalinity of the solution will be decreased by the formation of insoluble magnesium hydroxide and calcium sulfate.

$$Ca\,(OH)_2 + Mg\,SO_4 \rightleftarrows Mg(OH)_2 + Ca\,SO_4$$
$$pH\ 11\text{--}12 \qquad\qquad pH\ 5\text{--}7$$

The buffering action stabilizes the pH in the desired range and the ammonium is not present. The system is self-adjusting and good practical results are claimed.

Further deliming may take place by the addition of acid, which helps to clarify the grain surface by dissolving calcium hydroxide and preventing the formation of calcium carbonate. Lactic acid is often used for this purpose and is particularly desirable for the adjustment of pH in the production of vegetable leathers.

During the deliming process, the tanner may cut the skin with a knife and use an indicator, usually phenolphthalein or thymol blue, to determine the "lime streak" in the skin and thus measure the depth to which the deliming operation has proceeded. The extent of the differential swelling and the distortion of the fibers can also be observed. When the hide is in this condition care must be taken that excessive mechanical action is not used; the action of the drum may result in a rupture of the fibers and a looseness of the leather. After the skin has been thoroughly delimed, the stresses are no longer present and the skin can withstand mechanical action.

The deliming procedure also affects the solubilization of the fats. In liming, the free fatty acids of the hide fats are converted to calcium soaps of very limited solubility.

In the deliming operation the calcium component of the calcium soaps is solubilized and removed as complex calcium ammonium salts. Much of the fat removal will take place during the deliming and subsequent bating operations. Ammonium and sodium soaps are soluble in a pH range around 8. Since the bating operation is at a pH of 7 to 8 and the temperature is near 100° F, this part of the process is very important for fat removal. In the chapter on fatliquors we discuss emulsions and the solubilization of fats and oils.

With the high incidence of fats in the heavy cattle hides presently being worked by American tanneries, additional fat removal by the use of detergents may be required. The use of detergents is not simple because some detergents can be taken up by the skin, carried over to subsequent operations, and may affect leather quality. Nonionic detergents have been found to be most effective in a number of tannery applications, for they aid in the dispersion of fats and minimize the problems resulting from carryover.

The iron sulfides, which give the hide a faint blue color or dark blue stain, dissolve by the lowering of the pH in accordance with the law of mass action. There has been some work done on the use of chelating agents in deliming and bating operations for aid in the removal of iron and calcium salts. These treatments can be very effective, but proper choice of the chelating agents is critical to avoid interference with subsequent chrome tanning operations.

The swollen skins, prior to deliming, can be washed to remove surface filth and lime. In the hair-burning process for cattle hide sodium sulfide is absorbed into the skin, producing a strongly alkaline condition. Sulfides on the grain surface must be thoroughly washed out to avoid crackly leather out of tan. This operation requires washing with running water, sometimes up to an hour. The large quantity of water required and the sulfide it contains presents a significant effluent problem. Completion of the washing from a high sulfide lime is usually judged by means of the lead acetate test. A solution of basic lead acetate is dropped on the test section of the skin and the development of black lead sulfide is noted. The darkness of the spot and the speed with which it develops constitute a qualitative test for sulfide from which the tanner can judge the completeness of the washing procedure.

The practical bating operation takes place in a warm bath near pH 7 to 8. The presence of calcium salts is helpful in the activation of most bating enzymes, and deliming adjusts the pH for best enzyme action. The purpose of the deliming operation, therefore, is to bring about these changes gently and without damaging the skin. During deliming the temperature is usually raised gradually by washing with hot water. The temperature of the water added is about 100° F. High temperatures must be avoided to prevent damage from overheating the tender protein fibers. The ammonium salts also activate the enzyme, and the deliming and the bating procedure is continuous in most tanning operations.

BATING

Bating is a further step in the purification of the hide prior to tanning. The unwanted components consist of some of the protein degradation products, epidermis, hair, and the "scud" on the surface of the skin and in the hair follicle and pores. Some of the chemically resistant, fibrous protein must also be removed in order to obtain proper grain texture and leather softness. Bating refers to the action of enzymes of these components.

The study of enzymology is probably as old as the study of tanning itself. Bacterial action and enzymatic action are so closely related that until recently it was impossible to determine a clear distinction between the two.

Figure 1   Fleshing machine as used for side leather. Similar machines may be used for unhairing or scudding (removing fine hair and scud) using dull blades. Similar machines are made on a smaller frame for light leathers and small skins.

The biological functions of living cells are controlled by specific biological catalysts (enzymes). Many enzymes have been separated as crystalline products, but even these very complicated protein structures have not yet been isolated or identified.

Enzymes, even commercial enzymes, must be obtained from living cells. Commercial enzymes are obtained from four major sources: animals, plants, bacteria, and fungi. Two major examples of animal enzymes are rennin from calves' stomachs and trypsin (bating enzymes) from hogs' pancreas. Calf stomachs are of great value, with a market price of over a dollar each. They are dried, ground with wood, extracted, and the enzyme rennin recovered. Rennin is essential for the manufacturing of cheese. The hog pancreas is ground and carried on wood flour to produce some of the enzymes in bating. Plant enzymes are also of major commercial importance. One of the most notable being papain, from the sap of the papaya tree. This is one of the most widely sold commercial enzymes. It is used extensively in chill proofing beer, and it has many other commercial applications. Many of the industrial enzymes are of bacterial or fungal origin. The fungus is grown on a suitable substrate such as a malt mash (on trays). The culture is removed, and the enzymes are extracted and concentrated by means of solvents or salts as appropriate.

One of the first uses of enzymes in leather manufacture was in bating. In primitive tanning, bating was accomplished by puering the hides by the use of chicken or dog manure. In this system an infusion of the manure was made with water and added to the limed hide. Although the procedure was followed very closely by the bate master, the action of the enzyme was not easily followed; it was simultaneous with the deliming, and quality control was very difficult. The skill of the bate master at this point was more in judging the degree of plumpness of the skins than in recognizing the actual extent of enzymatic activity.

O. Röhm, a good scientist, thought that the best way to improve quality control was to use an enzyme preparation rather than an uncontrolled material such as manure. Through research he developed the use of an extraction of ground pancreas to obtain enzymes useful as bating materials. The ground-up pancreas, carried on wood flour along with buffering salts and deliming agents, was introduced to the leather industry as a prepared bate. This ingenious invention was important in the establishment of the Röhm and Haas Company, and the leather industry became the father of a great chemical company.

Enzyme materials have very strong catalytic action. The specific action of an enzyme may be to act on a particular protein under a specific set of conditions related to the sequence of amino acids in the proteins. An enzyme can be introduced into a system which will have a very strong action on one particular protein and yet no apparent action on another. Commercial enzyme preparations are more or less specific depending upon the purification. Particular enzyme preparations may act on a number of different proteins, but at different rates. The activity of the enzyme and the nature of the activity varies with temperature, pH, and the presence of added materials.

If we plot the activity of an enzyme as a function of pH, we obtain a bell-shaped curve with a definite peak at some optimum pH. As the temperature is raised there is also an increase in the activity of the enzyme up to the point where it is damaged or the action of the heat on the substrate is greater than the action of the heat on the enzyme.

At 10° F difference in temperature may have a tenfold change in the rate of activity of an enzymatic action. On the other hand, doubling the quantity of the bating materials or enzyme may not double the rate of the enzymatic activity, since the action of the enzyme on the protein depends not only on concentration but also on the ability of the enzyme to get to the reactive portions of the hide.

In practical bating the actual amount of enzyme used is very low, and conditions are very closely controlled. If an enzyme is not applied under the conditions recommended by the manufacturer, there may be a waste of bating material, and possibly poor quality leather will result. The bating operation and the quantity of bating enzyme used differ for different types of skin. Usually, in the bating of heavy hides, for sole leathers, or other vegetable tannages, the quantity of bating enzymes necessary and the extent of the bating action are very little, for the hide has been well limed and most of the protein degradation products have been removed.

The liming operation for side leather is not so extensive as that of sole leather, and a slightly stronger bating action is needed for it than for sole leather. In side upper leather the bating and deliming operations usualy take place simultaneously, with the pickle following before a full removal lime streak is obtained. In such bating systems only a small portion of the skin is under conditions where true enzymatic activity can be effective. The importance of bating enzymes in this application is minimal.

In the harder natured goatskins, higher concentrations of bating enzyme are used, and the bating proceeds for a much longer time. There are a number of commercial procedures in bating goatskins which involve bating overnight in a paddle.

The bating process is quite effective for the removal of fat from the skins through the adjustment of pH, temperatures and through a detergent action of the natural soaps. An additional effect claimed by some manufacturers is the action of the bating enzymes in the breakdown of the fat cells.

By the time the bating has proceeded to the point desired, the differential swelling effects have been overcome and the skin can stand the mechanical action. It is desirable to wash out the chemicals and the degradation products from the bating application and to lower the temperature to stop the action of the enzymes. Cold water is added and these skins are run, with a continual flow of water, through the drum or paddle until the wash becomes clear and the temperature is the desired level, usually about 65° F.

PICKLING

Pickling refers to the treatment of the hide with salt and acid to bring the skin to the desired pH for either preservation or tanning. At the end of the pickling operation the skin is theoretically a purified network of hide protein. The previous operations of curing, soaking, liming, deliming, and bating have all been concerned with the removal of unwanted components. On this hide fiber can now be built the chemical reactions which will produce the desired character of the leather. The pickling adjustment of the skin to the desired pH is the first of these constructive steps. The pH desired will depend upon the tannage to be used and the time between bating and the start of tanning. Provided a proper quantity of salt is used to control the swelling, the pure hide protein or unhaired skin can be brought to a very low pH (2.0 or lower) without having a significant breakdown of the skin due to acid hydrolysis.

For preservation by pickling, the skins may be treated with a solution containing one pound of salt per gallon of water. After tumbling in this solution long enough to prevent osmotic swelling (usually about ten minutes) the acid may be added and the pH will be adjusted by the absorption of the acid into the skin. At the lower pH the acid penetrates into the skin, but acid swelling is avoided since the salt has already penetrated ahead of the acid. From our discussion of osmotic swelling and acid absorption by hide protein (Ch. 4), we can see that it is much better to work with known quantities of acid than to try to adjust the skin to a particular pH. The buffering action and lack of true equilibrium condition is so great that gross errors could result here if the pH were to be adjusted strictly on the basis of pH.

In calculating the amount of salt necessary we should take into consideration not only the water that is in the solution around the skins or added with the pickling, but also the water that is carried within the skin itself. It is customary to maintain the salt concentration at approximately 5% sodium chloride (solution basis), but salt concentrations as low as 3% may be desired for some particular tannage. The tanner should not consider the quantity of salt from the point of view of expense. The fiber characteristics are controlled to a great extent by the pH, and the salt concentration as such is one of the most critical factors in the quality of leather to be produced.

In light skins where the pickling is conducted for preservation, the pickle action is allowed to reach equilibrium, usually overnight, and the skins absorb acid to a pH of approximately 2 or even lower. This strong acid condition will prevent the growth of bacteria, and provided sufficient salt is present, the acid will not damage the skin. The high salt, low pH condition

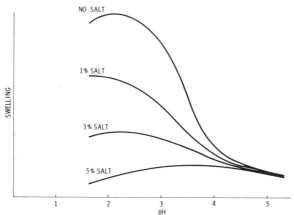

Figure 2   The amount of swelling of a hide protein can be controlled by salt in the solution. The more salt, the less swelling, although about 5% salt on a solution basis is a good practical upper limit. In calculating the weight of salt needed, particularly in a drum pickling process, the water with the bated stock should be taken into consideration.

will not necessarily prevent the development of molds, for many molds will grow primarily on animal fats. The presence of large quantities of fat in light skins, particularly sheep, requires disinfectants to prevent mold growth. Practical disinfectants include the chlorinated phenols, some florides, and specialty commercial products. Pickled skins may be preserved and held for over a year prior to further processing into leather.

For immediate processing, the pickling pH is usually less extreme; a pH of approximately 2.5 to 3 may be desired. Light skins, such as sheep, goat, and calf are usually pickled to equilibrium at the desired pH.

In the bate, pickle, and tan system for side leather the deliming, bating, pickling, and chrome tanning are sequential steps in a single batch process. In the interest of production efficiency this process is done in large drums (most common size is 10 feet x 10 feet for about 10,000 to 12,000 pounds of limed hide). The production rate forces as quick a sequence of operations as possible with consistent quality. Under such a system complete pickle equilibrium may not occur. The salt-acid balance and the rate and timing of the additions determine to a great extent the character of the leather produced in a chrome tannage.

A strong acid-salt pickle would be much too harsh for proper vegetable tannage, particularly of heavy leather, because of incompatibility with the vegetable tanning materials. A low pH would cause fast surface tanning and hardening of the leather. Sodium acetate or other materials are often added to pickle solutions to buffer at about pH 4 for the pickling of heavy leather hides. The balancing of the pickle for both vegetable and chrome tannages is the major determining factor for the rate of penetration of the tanning agents. There are two general rules to be followed in the pickling of hides and skins: (1) sufficient salt must be available to prevent acid swelling; (2) the quantity of acid used must be carefully determined for the particular formulation and must be added on a proper time schedule for consistent results. The timing is of particular importance in heavy leathers where continuous processes are used.

Recent work by the author and his associates has shown that weak organic acids may be used. In this regard, the weaker the acid is, the higher the rate of penetration of the pickle acid. Acetic acid penetrates faster than formic acid which penetrates faster than sulfuric acid. This is of practical value for the development of high speed bate, pickle and chrome tan systems which will be discussed under chrome tanning.

Depickling

Occasionally, the tanner buying pickled hides or skins will want to adjust the pH higher than that carried in the skins. To do this the skins should be covered with brine and water containing the proper amount of salt. Then the pH can be raised. This procedure is seldom satisfactory because the pH of the solution may have little relation to the quantity of the acid remaining in the skins. The equilibrium conditions are seldom established and the usual results of such a procedure are inconsistent quality and such side effects as poor grain, undertannage, and hard leather. If a depickling system is to be used, it should be at least overnight to equilibrate the pH.

Most pickling operations are based on the use of a sodium chloride-sulfuric acid system. For chrome tanning, some tanners prefer to use formic acid as a portion of the acid. This provides buffering and will aid in the masking of the chrome tan. The use of a buffered system or formic acid decreases the possibility of acid swelling, and a smoother, more uniform leather may result.

Another possible system is the use of the saltless pickle which has been proposed in Europe. This system involved pickling with organic acids, particularly sulfophthalic acid, to obtain a uniform chrome tannage. The sulfophthalic acid pickle has been given little attention in the United States where it is customary to pickle to a somewhat lower pH than is usual in Europe.

LOOKING AHEAD

The acid-binding capacity of hides has been investigated rather thoroughly from a scientific point of view. The practical effects of pickling are more empirically approached by the tanners. The pickle and degree of pickling are of great importance in determining the quality of the leather and the speed of the tannage.

Cattle hides will probably be sold in increasing numbers in the pickled state, which will cause new problems in pickle quality standardization for the tanner. Buffered pickles and possibly light pretannages may be incorporated into the pickling of hides and skins for preservation before marketing.

In order to overcome the variations in pickling and their effects on practical tannages, a complete review of the basic principles of pickling, from the points of view of both the tanners and the research chemists, is essential.

The use of pickling for maintaining preservation of hides has been well established for sheepskins, goatskins, and some others. An attempt was established by several persons, including the author, to establish pickled cattle hides as an accepted article of commerce. This, however, was not successful as a commercial venture. Certain segments of cattle hides, however, are sold in the pickle—namely, shoulders and bellies. Pickled bellies are quite common as a by-product of the vegetable tanning of heavy leathers.

REFERENCES

J. Am. Leather Chemists' Assoc., Supplement No. 2, Round Table Discussion (1952).

J. Am. Leather Chemist' Assoc., Supplement No. 3, Leather Making (1952).

J. Am. Leather Chemists' Assoc., Supplement No. 4, Leather Making (May 25, 1955).

McLaughlin, G. D., and Theis, E. R., "The Chemistry of Leather Manufacture," ACS Monograph 101, New York, Reinhold, 1945.

Pankhurst, K. G. A., "The Chemistry and Technology of Leather" (O'Flaherty, Roddy, and Lollar, ed.), Vol. 1, New York, Reinhold, 1956.

Stather, F., "Gerbereichemie und Gerbereitechnologie," Berlin, Akademie Verlag, 1948.

Stubbings, R., "The Chemistry and Technology of Leather"(O'Flaherty, Roddy, and Lollar, ed.), Vol. 1, New York, Reinhold, 1956.

Wilson, J.A., "The Chemistry of Leather Manufacture," Vol. 1, New York, Reinhold, 1928.

CHROME TANNING

8

Tanning may be defined as the treatment of hides and skins for preservation and conversion into useful articles of commerce. To define tanning in terms of preservation or chemical reaction alone is inadequate, since the quality of "leathering" is not considered. A definition based on physical properties associated with leather-like characteristics is not adequate for a scientific approach. In describing the criteria which define tannage, both aspects must be considered. Vegetable, aldehyde, oil, and most other tanning methods yield usable leathers from the tanning process alone. In chrome tanning, the reaction with the chromium salts gives a very stable hide fiber which is resistant to bacterial attack and to high temperatures. Without further processing however, chrome tanned leather does not have many of the qualities desired for useful articles. Chrome tanning must be used in conjunction with the additional processes of dyeing, fatliquoring, and perhaps vegetable re-tannage to produce usable leathers. The main advantages of chrome tanning are high speed, low cost, light color, and excellent preservation of the hide protein. Chrome tanning rapidly took its place in the commercial world shortly after its discovery and became the most common method of tanning light leathers and shoe upper leathers.

HISTORY

The discovery of chrome tanning is attributed to Knapp in 1858. The first commercial production of chrome leather is attributed to Augustus Schultz of New York in 1884. The original patents of Schultz were based on a two-bath tannage. In the two-bath system the skins are treated with an acid dichromate solution, and later in the process the dichromate is reduced in situ to the green trivalent state. The system, discussed more fully later in this chapter, had many drawbacks, but it had specific desirable leather-making qualities. The two-bath chrome tanning method has continued to be popular for some types of skins, though it has been considerably modified for modern production.

Commercial chrome tanning in the United States is done almost entirely by the one-bath process. This is based on the reaction between the hide and a trivalent chromium salt, usually a basic chromium sulfate. In the typical one-bath process the hides or skins are in a pickled state at a pH of 3 or lower when the chrome tanning materials are introduced. At these low pH values the affinity of the chrome tanning salt for the protein is moderate, allowing penetration of the chrome tanning agent into the hide. After proper penetration and initial absorption of some of the chrome salt have been achieved, the pH is raised. This brings about changes in both the chromium salts and the protein and causes a reaction between them. When the reaction is completed the leather is said to be full chrome tanned. In this condition it will stand immersion in boiling water. The chemistry of the chrome tanning process is complicated and involves several simultaneous, competing reactions. It is the balance of these reactions, achieved through control of the conditions of temperature, pH, and materials used during the chrome tannage that produces the quality of leather desired. The skill of the tanner in dealing with these factors makes possible consistently high quality leather.

CHEMISTRY OF THE CHROMIUM SALTS

Chromium tanning salts have a valence of $+3$; they are soluble in strong acids but will usually precipitate as chromium hydroxide, or hydrated chromium oxide, at pH above 4. They react with a number of organic materials to form colored soluble salts at higher pH values, and will precipitate soluble proteins.

Basicity concept

Chromium $(+3)$ in a solution has a strong attraction for hydroxyl ions (OH^-). The reaction of chromium with the OH may be written as a three-step reaction as it takes on the first, the second, and then the third hydroxyl group. The tendency of the first step of the reaction to take place is very strong, and even at pH 2 (a concentration of OH^- in the solution of 10^{-12}) the chromium will hold the first OH group. As the pH is raised (the concentration of OH^- increases) the second OH group enters into reaction with the chromium. This takes place between pH 3 and pH 4. Near pH 4 the third OH enters into the reaction. In order to complete the reaction with the third OH group it is necessary to raise the pH to about 8 or 9 and to bring the temperature of the solution to the boiling point.

The percent of the primary valence bonds of the chromium in solution accompanied by OH groups is called the basicity of the solution.

$$[Cr]^{+3} + OH^- \rightleftharpoons [Cr-OH]^{++} \qquad \text{33\% basicity}$$

pH 2.0 and below—approximately

$$[Cr-OH]^{++} + OH^- \rightleftharpoons [Cr(OH)_2]^+ \qquad \text{66\% basicity}$$

pH 2.0—pH 4.0—approximately

$$[Cr(OH)_2]^{++} + OH^- \longrightarrow [Cr(OH)_3]^0 \downarrow ppt \qquad \text{100\% basicity}$$

pH 4.0—pH 8 or 10 may require heat

The chromium salts used in chrome tanning usually have basicities between 33 and 45%.

Coordination complexes

The three main valences of the chromium atom are not its only attractive forces. In addition, there are three secondary forces or coordinate positions which can attract ions and molecules from the solution. Chromium has six coordinate positions, so we may draw the chromium atom in solution as indicated:

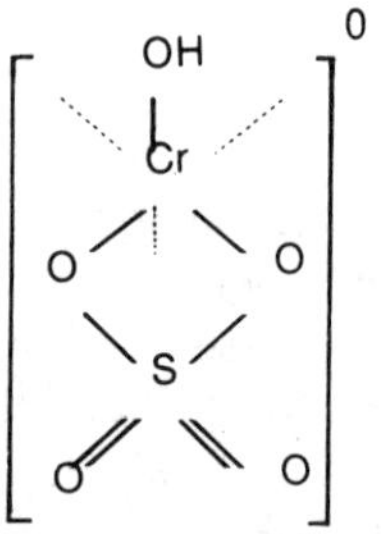

33% basicity (*chromium sulfate*)

66% sulfate acidity

Note: Will form mostly cationic complexes and $SO_4^=$ in solution.

These coordinate positions may contain water molecules, hydroxyl groups, or other materials in the complex. Any material that is capable of reacting or is in a coordinate position is called a ligand. A typical complex of basic chromium chloride could be written as indicated. As the term "basicity" refers to the association of the OH with the chromium, the term "acidity" is applied to the acid portion of the salt. The sum of the basicity and the acidity, expressed as percentage, for a given salt in solution totals 100, by definition. Thus

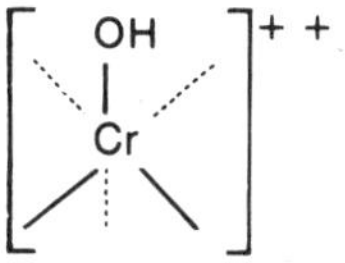

$$\begin{bmatrix} & OH & \\ & | & \\ & Cr & \\ \diagup & \vdots & \diagdown \end{bmatrix}^{++}$$ 2 Cl⁻

33% basicity

Cationic complex

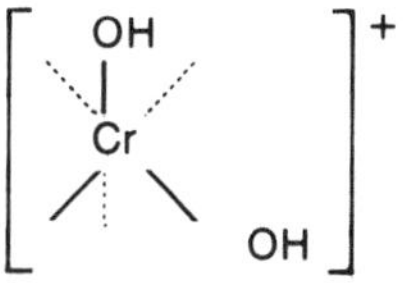

$$\begin{bmatrix} & OH & \\ & | & \\ & Cr & \\ \diagup & \vdots & \diagdown \\ & OH & \end{bmatrix}^{+}$$ Cl⁻

66% basicity

Cationic complex

$$\begin{bmatrix} & OH & \\ & | & \\ & Cr & \\ \diagup & \vdots & \diagdown \\ Cl & \vdots & Cl \end{bmatrix}^{0}$$

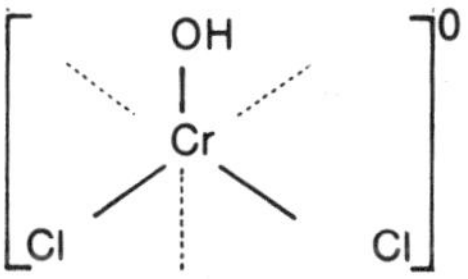

33% basicity

66% *chloride* acidity

Nonionic

Note: 33% basic *chromium chloride* will form cationic *chromium complex* and Cl⁻ in solution.

The acidity components of the salt may be associated with the chromium atom in a complex ion, or they may be in the solution.

Masking agents

It was indicated above that the chloride ion could either enter into the complex or remain in the solution. Chloride has little tendency to enter the chromium complex, but other materials may be strongly attracted into this complex. The sulfate ion, for example, is attracted in such a way that about one third of the primary valences of the chromium complex are usually occupied by sulfate.

The reactivity of a number of different organic materials with chromium in solution has been studied extensively. These materials can be used to control the chrome tanning process. Such materials are called masking agents because they modify (or mask) the chromium tanning reaction to give improved tannage characteristics.

The conditions of the chromium salts in the solution can be represented by this idealized picture. There are three stages of initial reaction represented.

(1) Solution: The basic chromium sulfate dissolves, and as the sulfate ionizes, a cationic chromium complex is formed.

(2) Masking with sodium formate: The formate ion in the solution forms a basic chromium formate complex with the displacement of some of the sulfate from the chromium complex.

(3) Fixation: The cationic chromium complexes react with the anionic carboxyl groups of the acid amino acids of the hide protein in the initial tanning reaction.

The formate ion has a greater affinity for the chromium complex than chloride or sulfate ions. Formate ions added to a solution containing the basic chromium sulfate will displace the sulfate and the formate will enter into the complex. Acetate ion has a still greater affinity than formate for chromium.

The charge on the complex is equal to the sum of the charges of the chromium and the coordinating ligands. The addition of negative charges into the chromium complex will decrease and neutralize the $+3$ charge stepwise, and finally the complex will become zero charged (nonionic) or negatively charged (anionic).

In the study of masking agents it was established in very early research that the anions could be listed in order of increasing affinity for the chromium tanning complex. This series, as proposed by Stiasny, is nitrate, chloride, sulfate, formate, acetate, sulfite, and oxalate, going from the least attractive (nitrate) to the most attractive (oxalate). The formate ion, which has been found to have the most desirable properties of stabilization of all the chrome tanning agents, results in a complex of particularly good tanning properties. Oxalate is a very strong complexing agent that will react with the chromium tanning agent to form a complex in which all six positions are occupied by oxalate, and in which the complex has a -3 charge. Complexes in which more than one position is occupied by a single complexing agent are called "chelate" complexes.

Solution

Masking

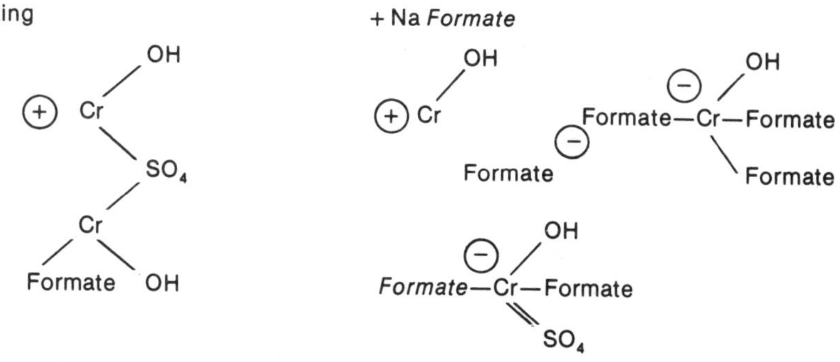
+ Na *Formate*
OH
Cr
SO_4
Cr
Formate
OH
Formate—Cr—Formate
Formate
Formate
Formate—Cr—Formate
SO_4

Fixation

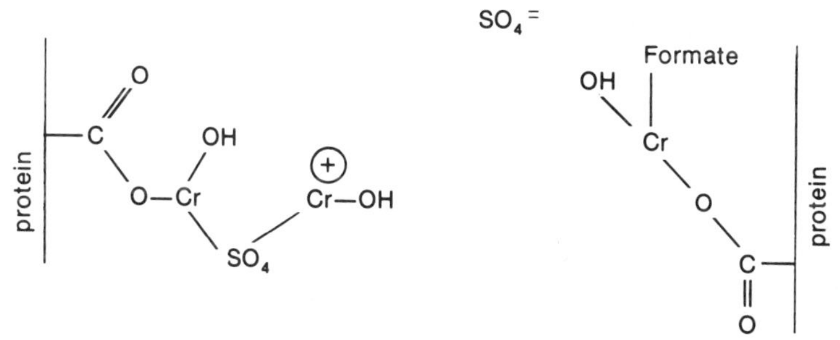
protein
O
C
OH
O—Cr
SO_4
Cr—OH
$SO_4^=$
Formate
OH
Cr
O
C
O
protein

Hide protein

The hide protein contains free carboxyl groups and other reactive sites which could, theoretically at least, form coordinate complexes with chromium salts. There is some disagreement among research workers as to the basis of the reaction of the protein. Most agree, however, that the initial tanning reaction is dominated by the affinity of the free carboxyl groups of aspartic and glutamic acids for the chromium salt in the solution. This simplification will be sufficient to solve most practical chrome tanning problems.

The hide protein can be considered as a coordinate ligand. The carboxyl group (COOH), when ionized as a carboxylic acid, will be attracted to the chromium tanning complex and a reaction will take place. When the carboxyl group is not ionized there is little attraction between the protein and the chromium. The fixation of the chromium by the protein increases with the increased ionization of the hide protein carboxyl group, and the fixation curve rises from practically zero at pH values less than 2 to a maximum at pH values of approximately 4. This strong shift over a relatively narrow pH range corresponds almost exactly to the ionization of the carboxyl groups of the protein.

There are, then, four competing reactions taking place simultaneously in chrome tanning. These are all reactions between coordinating ligands on the chromium complex. As conditions of pH, temperature, and concentration are adjusted, the relative dominance of each reaction can be controlled. We can consider the entire chrome tanning process to be a control of the extent of these reactions during the tannage. Practical tanning is a skillful balance of these conditions during the various stages of tanning. The four reactions are

(1) the reaction between the OH group and the chromium, i.e., the basicity

$$Cr^{+3} + OH^- \longrightarrow \begin{bmatrix} OH \\ | \\ Cr \end{bmatrix}^{+2}$$

(2) the reaction between the cation of the chromium compound and sulfate

$$\begin{bmatrix} OH \\ | \\ Cr \end{bmatrix}^{+2} + SO_4^= \longrightarrow \begin{bmatrix} OH \\ | \\ {}^+Cr \\ \diagdown \\ SO_4^- \end{bmatrix}^0$$

(3) the reactivity of masking agents, such as formate

$$\begin{bmatrix} OH \\ | \\ Cr^+ \\ \diagup \\ SO_4^- \end{bmatrix}^0 + Formate^- \longrightarrow \begin{bmatrix} OH \\ | \\ Cr \\ \diagup \\ Formate \end{bmatrix}^+ + SO_4^=$$

(4) the reactivity of the hide protein

$$\begin{bmatrix} OH \\ | \\ Cr \\ \diagup \\ Formate \end{bmatrix}^+ + {}^-OOC\!-\!\!\Big| \longrightarrow \begin{array}{c} OH \\ | \\ Cr \\ \diagup \quad \diagdown \\ Formate \quad OOC\!- \end{array}$$

Hide protein

At low pH the concentration of OH $^-$ in the solution is low; the basicity of the chromium is also low. The first reaction is forced to the right as the pH is increased. The coordination of the sulfate is not strongly affected by pH, and the sulfate will be present in the complex at low pH. This reaction is not directly affected by pH since $SO_4=$ is a strong ion and remains reactive with the protein over the entire pH range of practical tanning. The masking agents may be coordinated with the chromium complex at low pH values, provided they are sufficiently ionized. The coordination of a weak organic acid (masking agent) with the chromium complex is dependent both on the nature of the acid and on the pH. Higher pH favors increase reactivity. The hide protein, since its ionization is repressed at the low pH values, has little reactivity with chromium. The carboxyl groups of the protein react in a manner similar to weak acids but are more affected by pH changes.

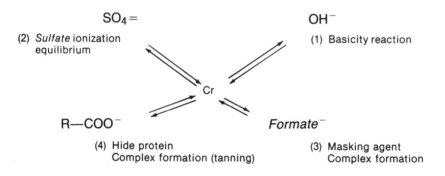

As the pH of the tannage is raised, the basicity of the chromium complex is increased (more OH enters the complex). The affinity for coordination of the sulfate remains the same. The masking agent remains coordinated with the chromium at this stage of the reaction. The reactivity of the protein is greatly increased with increased pH, and the initial tanning reaction is accomplished.

At the end of the basification the basicity is high and the sulfate may be partially displaced from the complex. The masking agents may be displaced as the hide protein gains in its affinity for the chromium tanning compound. This linkage between adjoining chains of the protein may be completed and a cross linking reaction will result. With the increase in basicity, two chromium compounds may be associated with one another through secondary attractive forces to the OH group in the manner indicated:

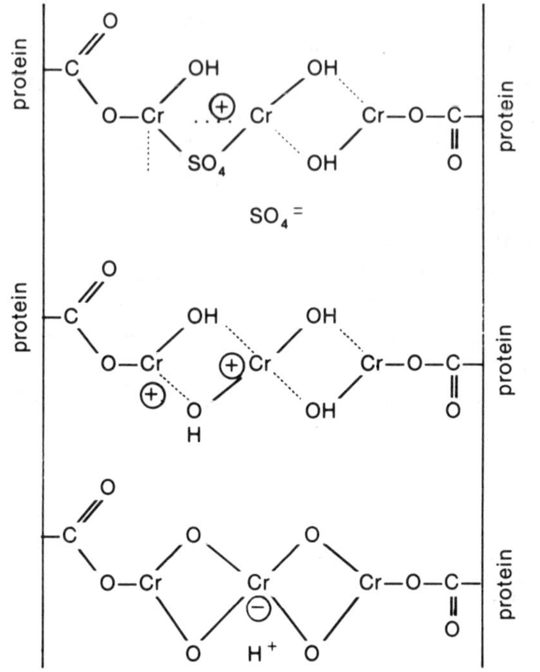

Simplified tanning actions:
* (a) crosslinking
** (b) basification—olation
*** (c) oxolation

Thus, the stages in the cross linkage of a chrome tannage are as follows:

(a) The chrome complexes have reacted with the protein carboxyl groups.

(b) As the pH of the tannage is increased, the sulfate associated with the chromium becomes displaced by the hydroxyl. The hydroxyl groups become shared by chromium atoms through olation.

(c) Upon drying, the tannage becomes more stable as the complex gives up hydrogen ions and oxolation results.

As the tannage proceeds at the higher basicities, the size of the complex increases, permitting this cross linking. This reaction results in a full tannage and a high shrinkage temperature.

Practical concepts of chrome tanning

Considering the principles outlined above, the factors affecting practical chrome tannage are pH, basicity, the nature of the complex, the concentration, the temperature, and the time.

Basicity. Chrome tanning agents are described by their basicity. The higher the basicity, the larger will be the complex in the solution. The association of two or more chromium atoms through common sharing of hydroxyl groups is called olation. Basicity is intimately related to pH. However, as the reaction between the OH in solution and the chrome tanning complex is not instantaneously established, a change in pH does not result in an immediate, new basicity.

pH. The pH of the chrome tanning bath is most important in that the higher the pH, the greater the tendency of the protein to react. If the pH is increased too rapidly or to too high a point, precipitation of some of the chrome tanning agent in the solution may result. It is necessary, therefore, to adjust the pH carefully during chrome tannage and to maintain proper conditions throughout the tannage.

Temperature. The higher the temperature, the more quickly the reactions take place. At higher temperatures there is greater fixation of the chrome tanning compound by the hide protein and greater olation of the chrome complexes. The hide protein is also affected by high temperatures. Differential swelling effects, uneven tannages, and drawn grain can be caused by excessively high temperature at the early stages of the tannage. Most chrome tannages start at low temperatures.

Time. The chrome tanning process and the establishment of new complexes, new basicities, olation, and masked complexes is not an instantaneous reaction. The rate of each reaction changes with conditions of pH and temperature. Proper control of consistent quality chrome tannage requires adherence to an established time-pH temperature relationship.

Concentration. At high concentrations more of the ligands in the solution will be associated with the chromium compound. The basicity of the complex may also be slightly lower as a result of somewhat lower pH values. The concentration and liquor ratio maintained in a chrome tanning process should be kept constant from one batch to another.

Stratigraphic analysis. In the study of chrome tanning it has been found that it is advantageous to determine, by analysis, the location of the chrome in the leather. In order to do this, a sample of the leather is cut in layers from grain to flesh. A chemical analysis is made on each of these layers for chromium and high protein. Although this is a tedious process, the information obtained is very valuable, and in the development work this "stratigraphic" technique is of great value.

The data in the figure illustrate the research done by Oshugi and Thorstensen, studying the rate of penetration of pickle acids in a bath containing chromium salts. The technique used in this study was to take samples of hide cut with a cork borer during the tannage. As the tannage proceeded, the samples were sliced into layers from grain to flesh, and stratigraphic analyses were performed. In doing the stratigraphic analyses, the slices were placed in water and the pH of the solution measured. The samples were then digested in a usual manner for Kjeldahl analysis, and a small portion of the Kjeldahl digested material was used for analysis of chromium using a colormetric technique. It was possible, therefore, to obtain the chromium content as well as the pH on each layer, at various times during the tannage.

There is evidence that the acid penetrates both from the grain and flesh side. There is a dramatic difference in the pH from the edges of the skin toward the center. In the center, there is a relatively high pH due to the presence of lime. The chromium salts are soluble in the acid solutions and will penetrate along with the pickle acid as the pickle acid penetrates into the skin. It is, of course, not possible to have any significant penetration of the chromium beyond the point where the chromium is soluble. Therefore, the chromium penetrates up to a point where the pH is about 3.5. This is the pH of a high fixation of the chrome tannage.

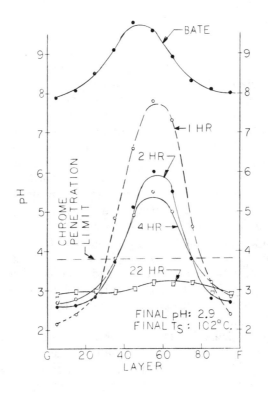

Figure 1   Stratigraphic pH of pickle-tanned stock. 1% Sulfuric Acid; 1.5% Cr_2O_3 as 33% Basic Chromium Sulfate; 5% NaCl.

From data of this type it can be seen why, under certain conditions of tannage, the amount of chromium fixed in the center of the skin may be higher than on the grain or flesh surfaces. Grain and flesh, having relatively low pH, would have low fixation of chromium tanning compounds, whereas the centers of the skin may have high fixation. These curves illustrate the changing conditions in the hide during a simultaneous pickling and tanning operation. The concentration and composition of the tanning salts, the pH, and the fixation of the tanning materials are all interrelated.

Practical examples. Let us consider a tannage in which the hides have attained an equilibrium pickle at pH 3. The chrome tanning agent (33% basic chromium sulfate) is added, also at pH 3. The chromium salt will enter into the skin and will be fixed upon the grain and flesh layers. As the tannage proceeds, more tanning materials are deposited in the grain and flesh, and less will be deposited in the center, in accordance with the normal diffusion pattern. The stratigraphic analysis, therefore, will have a dip in the center in accordance with the curves indicated. Upon basification, more fixation of the tanning agent will take place and the curve will be established at a higher level. The least effective area of chrome tannage, in this case, will be the center of the skin, and this section will have the lowest shrink temperature.

Let us consider now a system in which the leather has a pH of 2 on the surface and in which there remains in the leather some alkalinity from the bating process. The center of the skin has a pH of approximately 5–6. The acid of the pickle is still diffusing into the skin when the chrome is applied. The skin in the grain and flesh layers will have little affinity for the chrome tanning salt, and yet the chrome tanning agent will be strongly attracted to the center of the skin. When this condition prevails there may be a quick penetration of the chromium toward the center, and a dark green streak will be visible in the center of the skin. Shortly thereafter the pH of the center of the skin will approach equilibrium with the rest of the solution and a pH of about 3 may prevail through. With this particular tanning system the center of the skin, having tanned first, may contain more chromium than the grain and flesh layers. The leather in this case is tanned from the inside out. By proper balancing of the acid-base relationship at the time of the addition of the chrome tanning agent, the hides may have a slight lime streak or a slight alkalinity in the very heaviest sections at the initial stages of the tannage, and yet have reached an equilibrium pickle in the light belly areas. As a result of the inside-out tannage in the heavier sections, the problem of obtaining complete penetration of the chrome tanning agent is partially compensated.

It is a common practice in bate, pickle, and tan systems as used in the United States to start chrome tannage before equilibrium conditions of pH have been achieved in the pickle. The center of the skin may have a pH as high as 8 or 9 and may actually show a red color to a phenolphthalein indicator at the time of application of the chrome tanning salt. The pickle solution outside the skin may have a pH of about 2.5. The chrome tanning agent will be strongly attracted to the center of the skin. As the chromium complex approaches the center of the skin it is so strongly fixed that complete penetration of the tanning complex is not achieved. If this is the case, a cut on the leather will indicate five distinct color areas from grain to flesh: a light blue on the grain, a dark blue streak, a white streak in the center, another dark blue streak, and then a light blue on the flesh side. The chrome tanning agent penetrating the leather was completely fixed prior to reaching the center of the skin. As the tannage proceeds and more equilibrium conditions are established throughout the skin with respect to pH, this difficulty will be overcome and a uniform, full tannage will result.

The three examples above illustrate the dynamic nature of the chrome tanning system and the necessity for balancing pH of the pickle with the chrome tan.

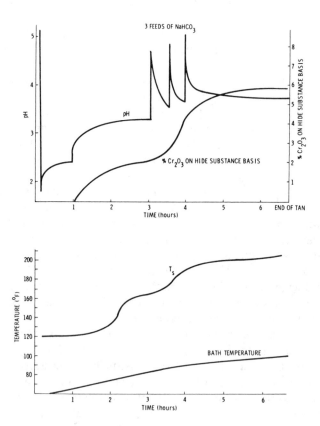

Figure 2   The relationships between pH, temperature, chrome take up and shrink temperature in a drum chrome tannage.

The set of curves shown below illustrates the relationship between pH of the solution, chrome take-up, shrink temperature, and the temperature of the tanning bath. At the start the pickle acid is added, lowering the pH. One hour later the chrome tanning salt is added. The curve of pH shows the take-up of acid and chrome by the skin. At three hours the bicarbonate additions are made to basify the tannage. The pH of the solution rises sharply with the addition but quickly falls as the reaction continues. The chrome fixation and shrinkage temperature are related and indicate the extent of the tannage. The bath temperature follows a steady rise due to the mechanical action. The bath temperature will eventually come to equilibrium at about 115–125° F, depending on mechanical conditions.

MASKING AGENTS

The sensitivity of the tannage to pH variations can be greatly decreased by the use of masking agents. These are additives to the chrome tanning bath which partially stabilize the complex and which are applied at basicities of about 33–45%. It is desirable to have a tanning complex that has less affinity for the protein and which will therefore be less sensitive to the pH variations outlined above. The addition of sodium formate or formic acid will result in a chromium formate complex accompanied by a corresponding displacement of the sulfate from the complex. The longer the contact of the formate with the chromium and the higher the temperature at which they are together, or the higher the mole ratio of the formate, the greater will be the tendency for the complex to develop. Some tanners prefer to obtain the chromium formate complex before adding the chrome tanning agent to the tanning bath; others prefer to have the formate present in the pickle as formic acid; still others add the sodium formate or calcium formate to the drum with the chrome tanning agent. Whatever system is used, consistency is necessary for quality production.

To compensate for the greater stability of the chromium formate tanning complex it is necessary to tan at a slightly higher pH than in an unmasked tannage. Other masking agents have been used in practical tanning. These include phthalate, sulfite, oxalate, and acetate. Each of these has its own special characteristics. The use of formate masking is by far the most common commercial method. Formate tannages, in general, are neutralized at the end of the tannage to pH of approximately 4, whereas basic chromium sulfate tannages are usually neutralized to pH of about 3.5. The pH at the end of the chrome tannage is one of the most critical control points in determining the quality of leather produced. The individual tanner must determine this value experimentally in order to obtain a leather which will be compatible with his entire production procedure and which gives the desired results.

The fixation of chrome tanning material as related to masking and pH can be seen in the figure below. This data is compiled from various laboratory studies by the author. The fixation of unmasked chrome tanning material by the hide is very sensitive to pH. Above a critical point precipitation takes place and fixation is sharply decreased. With formate masking, the reaction is less pH dependent. Higher pH is needed for optimum fixation, and precipitation takes place at a higher pH. With acetate masking, fixation is decreased; the pH at the precipitation point is higher. The decrease in fixation of chrome is so drastic as to greatly decrease the effectiveness of the tannage.

SHRINKAGE TEMPERATURE

The most common method of measuring shrink temperature is to clamp the sample between two supports, immerse it in a heating medium (usually water or a glycerin-water mixture), and raise the temperature. At the shrink temperature the sample will decrease in length. This decrease is indicated mechanically. The data are easy to obtain and are significant. The reaction between the chrome tanning agent and the hide protein will result in stabilizing the fiber by cross linking. The degree of stabilization can be most conveniently measured by the shrink temperature. In taking shrink temperatures it is necessary that the technique followed be the same from one test to the next in order to obtain consistent results. If the sample is taken from the tanning bath, placed in the shrink meter, and the temperature raised very slowly, there will be a tanning action during the time of the shrink temperature measurement. The shrink temperature in this case will be higher than if the sample were heated rapidly during the measurement.

In deciding on the proper shrink temperature for a chrome tanning process, a tanner must take into consideration the type of leather he wants to

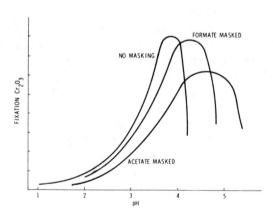

Figure 3   The effect of masking on the fixation pH characteristic of chrome tanning.

make and the nature of his subsequent operation. Some tanners prefer to use a slack tan. Under this system the chrome tannage is considered complete when the shrink temperature is a few degrees below boiling. A shrink temperature of 202–207° F (94–97° C) is considered ideal by some tanners. Slack tanned leather of this type may have a smoother grain and may be slightly more "let out" than leather which has had a full chrome tan and will stand a boil. In order to get the shrink temperature much above this point, it is necessary to raise the pH or temperature, or to tan longer. In all three of these cases a loose or soft leather may result. Subsequent chrome retannage or vegetable tannage will result in a full tannage.

The leather may be removed from the tan bath with a shrink temperature of about 205° F (96° C), yet after being piled on a pallet for a day or two the shrink temperature may be found to be high enough to stand a full boil. Other tanners disagree entirely and prefer that the leather, upon removal from the chrome tanning bath, be fully chrome tanned, i.e., it should stand a full boil, or, as some specify, a three-minute boil. In other words, a leather piece cut from the hide of the chrome tannage should be placed directly into boiling water for three minutes without any shrinkage. This is a very strenuous test, but it certainly indicates a complete tannage.

If the leather is brought to a full boil or a full chrome tannage under properly controlled conditions and a tight break is obtained, the full tannage will certainly permit the making of a wide variety of leathers with a minimum amount of lubrication in subsequent operations. In talking with practical tanners on this point the author has found that some of them, who operate tanneries with a reputation for high quality American leather, insist that a full boil shrink temperature is essential for consistent quality. At the same time, other tanners with equally good reputations believe that a slack tannage is best.

TECHNIQUES OF PREPARING CHROME TANNING SALTS

Chromium is obtained from the ferrichrome ores, i.e., naturally occurring mineral deposits containing the oxides of both iron and chromium. Upon refinement of these ores with alkali in high temperature kilns, sodium dichromate is formed. The dichromate is then separated from the iron by water. For chrome tanning, trivalent chromium is used, and therefore the chromium must be reduced from the +6 (dichromate) state to the +3 (basic chromium salts) state. Reduction of the dichromate can be accomplished with any reducing agent which will not, upon oxidation, form compounds that interfere with the chrome tanning process. Ideally, a solution of sodium dichromate can be treated with sulfur dioxide and water to produce a basic chromium sulfate in accordance with the following reaction:

$$Na_2CR_2O_7 + 3SO_2 + H_2O \longrightarrow 2Cr\ OH\ SO_4 + Na_2SO_4$$

This method was strongly recommended by Wilson for consistent quality production. The method is costly, however, and in most chrome tanning systems sugar is used for reduction of dichromate. In this case the dichromate is dissolved in water, sulfuric acid is added, and a solution of glucose or sugar is added slowly. The reduction of the dichromate takes place in accordance with the following equation:

$$8Na_2Cr_2O_7 + 24H_2SO_4 + C_{12}H_{22}O_{11} \longrightarrow 16Cr\ OH\ SO_4 + 8Na_2SO_4 + 27H_2O + 12CO_2$$

An excess of sugar is necessary since the oxidation is not complete and some organic oxidation products remain in the solution.

During the initial stages of the reaction almost all of the sugar is converted directly to carbon dioxide and water, since the solution is strongly oxidizing and strongly acid. Near the end of the process, however, very little of the sugar is oxidized and unreacted sugar may remain in the tan bath. Some tanners prefer to use a low temperature chrome reduction in order to hold some of the organic materials of oxidation to use as masking agents in the chrome tanning bath. Chrome tanning salts may be stored for a day or two prior to use in order to form a consistent quality complex for the tanning. The usual trend in American tanneries today is toward high temperature reduc-

Figure 4   In the preparation of chrome tanning liquors by sugar reduction, equipment such as this is used. The large tank in the foreground is the reactor containing the sulfuric acid and bichromate. This tank is lead lined and air-agitated. The smaller tank in the background contains the corn syrup. (*Courtesy John J. Riley Leather Co.*)

tion of the chrome liquor and the addition of masking agents as desired, rather than to rely upon the oxidation productions of the sugar.

PRACTICAL CHROME TANNING EQUIPMENT

Paddle tanning

Chrome tannage can be conducted in paddles of the same type used in the beamhouse operation. Paddle tannage employs liquor ratios of approximately four parts of solution to one part of hide, and under these dilute solution conditions the tannage proceeds more slowly than it would in a drum tannage. Temperatures during paddle tannage are lower than in drum tannage and have some advantage in the saving of materials, since the solutions can be re-used provided proper control techniques are employed. The paddle tannage process, however, has the drawback of being slow, and the labor of removing the leather from the paddles is very difficult and costly. The modern installation of chrome tanning processes in the United States is drum tannage or hide processors.

Drum tanning

Drum tannage in the United States involves a bate, pickle, and tan operation. In side upper leather, the sides are introduced into the drum in the limed condition. The sides are bated, pickled, and the tanning salts added (before pickle equilibrium is established). The condition of a light lime streak or light alkalinity in the center is the normal state of production. Using this system, usually with formate masking, the tannage can be completed in a cycle of approximately six to eight hours. Some tanners feel that the difference between a six or eight hour cycle and a twenty-four hour cycle on a tanning drum is very little, since normal labor and plant organization is best adapted to the latter. In this case the bate and pickle operation may be conducted and the skins reach near equilibrium pH prior to the introduction of the chrome tanning salt.

In the interest of efficiency the American tanning industry is turning more and more toward large equipment for chrome tanning. Drums 10 feet × 10 feet with a design capacity of 10,000–12,000 pounds of limed stock for bate, pickle, and tan operations are quite common. In some instances larger units are being used with capacities approaching 17,000 or 18,000 pounds. In addition to large capacity, some of these drums are being equipped with automatic feed equipment which will permit programming of the tanning operation on a completely automatic basis. Through the use of feed tanks, controlled output pumps, and controlled temperatures and pressures, these factors can be built into a card or taped program and the drum can be operated completely automatically. The drums are loaded either through

Figure 5 Large hide processing drums. Many tanneries have adopted very large drums for the earlier processes. In this case a drum capable of taking over 20,000 lbs. of hides is used for soaking, unhairing, bating, pickling, and tanning as an interrelated series without removal of the hides from the drum. The drum is mounted high, fed overhead, and dumps into boxes seen behind the operators. From here, the hides will go to the wringing machine and then onto a conveyor through the rotating knife that will cut the hide into two sides. (*Courtesy of Granite State Tanning Co.*)

the top, or the skins are placed in from a mezzanine above the door through a high lift-type fork lift truck. The drums are mounted high enough so that the tanned skins can be dropped out of the drum into a cart, thus eliminating the hand labor of pulling the drum. By dropping the skins into the cart, the drum is freed for the next batch and the operation is that much faster. As part of the automatic equipment there may also be a strain gauge in the support system so that the weight of hides and liquors in the drums can be measured while the skins are in the drum. In addition to wood, drums for the tanning industry have been made of stainless steel and of fiberglass. In normal tannery operation a wooden drum can be expected to last from ten to twenty years. Wood is an excellent material for construction of tannery drums, paddles, and tanks.

Dry tanning

Once the leather is pickled, preferably near equilibrium (uniform pH), the pickle solution can be drained and the chrome tan added in a solid state. Under these circumstances, as the chrome tanning agent is dissolved by the solution associated with the pickled hides, the complexes, upon first dissolving, are anionic and nonionic. As the ligands ionize from the chromium complex, the charge on the complex shifts from anionic to nonionic to cationic, and the tannage takes place. The high concentration of chromium salts in the solution and the salt in the residual pickle liquor is sufficient to prevent undue swelling of the skins during this tannage. The very low liquor ratio results in a greater temperature rise, due to the mechanical action, during the tannage. Tannage proceeds fairly rapidly, and usually within two hours after the introduction of the chrome tanning agent the solution can be neutralized by addition of sodium bicarbonate or any other appropriate alkali. Basification under the dry tannage system is often conducted by introducing the bicarbonate in a single feed. Undoubtedly, the high concentration of salts in the tan bath is also a contributing factor.

Weak acid chrome tanning

Recent studies by the author and his associates have shown that the rate of penetration of pickle acids is dependent upon the type of acids used. Through the stratigraphic technique it was determined that acetic acid penetrated faster than formic acid which penetrated faster than sulfuric acid. This discovery led to the development of a fast self-basifying chrome tannage.

In this system the hides are bated in a drum in the usual way. The hides then have residual lime in the center, particularly in the heavy sections of the hide (neck and bend areas). The pickle acid is mixed with the chrome tanning liquor to make a concentrated solution. This solution is then added to the drained bated hides in a single feed. The composition of the chrome tan feed is approximately as follows:

Chromic oxide as 33% basic chromium sulfate—15% on solution weight—1½% on limed hide weight—formic acid 10%–15% on solution weight—1-1½% on limed hide weight.

The solution liquor ratio is then 10% on limed hide weight. As in a dry tannage, the concentration of the solution is sufficient to prevent swelling. The pH of the liquor is low enough (less than 3.0) to prevent excessive surface tannage. The chrome tanning salts will penetrate the hide at the same rate as the pickle acid. Upon complete penetration, the pH of the solution should be between 3.5 and 4.0. The liquor can be drained and recycled to decrease pollution. No bicarbonate need be added to adjust the pH to the desired level. Dilution of the tan liquors should be avoided. The balance of the acid base components of the system and the masking effect of the formic acid will result in a fast smooth tannage. The time in the drum in contact with the chrome solutions should be about 4 hours, more or less, depending upon equipment, types of hides, etc.

Recycling of chrome tanning materials

Recycling of chrome tanning solutions has several advantages. First, there is the matter of cost saving. Second is a requirement to decrease the quantity of chromium in the effluent discharge. The decrease in the chromium discharge is mandated by the pollution regulations in almost every country. The recycling of chrome tanning solutions illustrates several interesting concepts in chrome tanning.

When paddle tannages were popular, recycling of the chrome tan was a matter of course. Each individual tan bath, on completion of tannage, was left in the paddle and the tanned hides pulled from spent solutions. Since the total volume of tanning solution to the weight of the hides being tanned was about 3 or 4 to 1, the chromium salts left in the spent solution were considerable. The value of such salts made the recycling of the solutions quite necessary.

For recycling, the salt content of the tan bath was built up to the desired level by the addition of solid salt, and additional pickle acid was added to the bath. This could be done with a chemical control system for each individual bath for close control, or could be done on the basis of averages, if the production of the tannery was quite stable. The bated hides, usually with some lime streak, were introduced into the spent chrome liquor containing the salt and the pickle acid. As the pickle acid penetrated, the chromium salts continued to penetrate with the pickle acid. In the chrome paddle tannages, approximately two-thirds of the chromium salt would be taken up by the hide. After a short time in the pickle, the additional chrome tanning salt was added and the tannage proceeded in the normal manner.

With the continued reuse of the paddle and reuse of the solution, there is gradual build-up of sulfates which equalize to a uniform level. The balance of materials lost with the solution, by spillage and being removed with the tanned hide, became equalized by the same addition of materials coming in the new tanning cycle. The total quantity of chromium salts lost to the drain is only a matter of spillage. The paddle system would, therefore, decrease the total amount of chrome going down the drain and would assure a good uniform quality of leather. Although the paddle tannage system for cattle hides is considered obsolete due to high labor costs, it may be the most practical and efficient system for some types of leather in relatively low labor cost areas.

Recycling in drum tannages

In a drum tannage, recycling is not as easy, since there is a dilution of the tanning solution during the neutralization phase of the tannage. The volume, therefore, of the total tanning liquor used becomes greater than the volume of water that is going to be used in the next pickling cycle.

The floor drainings and the draining from the drum may be screened and placed in tanks and simply used on a recycle basis. Because of the dilution, of course, there is a continual expansion in volume which will mean that some of the chrome solution must be dumped, or the spent chromium salts recovered by another means. The most common way of recovering spent chromium salts under these circumstances is by precipitation. The pH may be raised to the precipitation point of the chromium salts. The chromium salts precipitate, as a hydrated chromium oxide, leaving a supernate liquid of essentially chrome-free solution. The chrome-free solution may be dumped.

Proper conditions, and the addition of the materials, may be carefully controlled at this stage in order to avoid an excess of volume of the chromium hydroxide precipitate. The precipitation of the chromium hydroxide is complicated by the amphoteric nature of chromium. At pHs below 5–6 the chromium is predominatly cationic with the chromium accepting hydroxyl ions. At pHs above 7–8 the chromium hydroxide acts as a weak acid and becomes an anion. Only in a very narrow pH range will the precipitation have a minimum charge and minimum hydration.

Precipitation with sodium hydroxide or sodium carbonate will result in a gelatinous precipitate. If lime (calcium hydroxide) is used, a better precipitation will result. Very good success has been attained by the precipitation with magnesium hydroxide. Magnesium hydroxide is less soluble than calcium hydroxide and as a result, the hydroxyl ion concentration is less and the pH lower. At pH 5–7 the magnesium hydroxide will come into equilib-

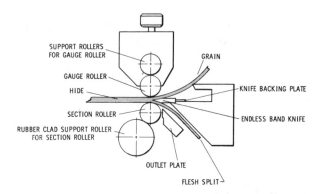

SUPPORT ROLLERS
FOR GAUGE ROLLER

GRAIN

GAUGE ROLLER

HIDE

KNIFE BACKING PLATE

SECTION ROLLER

ENDLESS BAND KNIFE

RUBBER CLAD SUPPORT ROLLER
FOR SECTION ROLLER

OUTLET PLATE

FLESH SPLIT

Figure 6 Splitting machine. The splitting machine is a precision band knife that will level the skin to a uniform thickness. By its use a grain layer or a split can be obtained for shoe upper leather. For upholstery leather, where very light leathers are needed from large, thick hides, several layers may be cut. The belt knife of the splitting machine is a precision-ground loop knife kept sharp by two properly positioned grinding wheels. The position of the knife is accurately controlled by means of settings by the operator. In the more modern machines several of these settings may be coupled together to compensate for the wear on the blade.

Above the knife blade is a guide cylinder which can be adjusted accurately with relation to its position to the blade. The distance between this cylinder and the blade controls the thickness of the cut. The leather is held against the upper cylinder and the thickness of the grain layer is kept constant by the position of the blade. The thickness can be adjusted to any desired degree in accordance with the type of leather being made.

Since the leather is not of uniform thickness, some parts of the hide must be cut away more than others. This difference in thickness is compensated for by a group of floating wheels supporting the hide against the top cylinder. A series of brass wheels supported by a rubber roll and running on a common shaft have sufficient clearance so that they can move up and down independently of one another. As the hide is fed through, the thicker sections force the supporting wheels down further than the thinner sections while the skin is still held tightly against the upper reference roll, or bolster. As a result, grain layer of uniform thickness is obtained. The accuracy of the grain layer during the splitting is within one half ounce (1 oz = 1/64 in.) on heavy leathers in the hands of a good operator. The split, after it is removed, is trimmed and may be sent to the splitting machine again for leveling. Note: The thickness of leather is measured in ounces in the English system 1 oz = 1/64 in. In the metric system thickness is measured in mm. 1 oz = .4 mm 1 mm = 2.5 oz.

Splitting machines come in several sizes, depending upon the size of the leather to be split. In a shoe factory, splitting machines with a cutting area as small as one foot are used to adjust the thickness of some shoe pieces. In calfskin and side leather plants the cutting area on a splitting machine is usually 60 or 72 inches. In upholstery leather plants the width of the splitting machine may be as much as 120 inches. (*Courtesy Turner Tanning Machinery Division of United Shoe Machinery.*)

Figure 6 (cont.)

Figure 7 Shaving machine. This machine is designed along the lines of the modern feedback shaving machines in that the bolster is mounted high and the entire width of the skin is shaved in one sweep. It is a feedback machine that allows the shaving of the entire side in two feeds, front and back. Machines are available up to sixty inches wide with feed speeds of up to sixty feet per minute. Machines of this type permit much faster and more precise shaving than is possible with smaller units. (*Courtesy Chas. H. Stehling Co.*)

rium with the chromium so the precipitate will be formed with minimum hydration. Once the precipitate has been formed, it can be redissolved by the addition of acid or may be introduced into the chrome reduction tank, where it will be dissolved by acid and reintroduced into the chrome tanning.

The saving of chromium by recycling methods may be well worthwhile from a cost of materials perspective. Care must be taken, however, that the quality of the leather be maintained, and this can only be done by constant chemical control, chemical analysis, and good attention to detail.

By chrome recycling on the main chrome tannage, the quantity of chromium to be discharged is usually low enough so that effluent regulations can be met. During the coloring and fatliquoring, additional chromium may be discharged into the solution as particulate hide material is rubbed off the leather or a solubilization of some of the chromium salts takes place. This amount of chromium, however, is usually not objectionable to the discharge system of the tannery, and certainly is not of sufficient value to warrant recovery.

Two-bath chrome tanning

A two-bath chrome tannage, as outlined earlier, is based on the penetration of the skin by dichromate which, in turn, is reduced by the additon of sodium thiosulfate during the tannage. This results in an in situ formation of trivalent chromium and fixation of this chromium by the skin. Two-bath tannage also results in the formation of sulfur from the breakdown of sodium thiosulfate. The formation of sulfur inside the skin has a filling action and results in a certain roundness, a fullness, and tightness of grain that is desirable. Two-bath chrome tannage is costly in materials, labor, and time. The two-bath chrome tannage has disappeared from practical application in side upper leather production in the United States but is still used in the production of some light leathers, such as glazed kid. The colloidal sulfur aspect of the use of sodium thiosulfate has led some tanners to incorporate thiosulfate into the neutralization process as a part of their normal one-bath chrome tannage.

Aluminum tanning

Aluminum salts have greater affinity than chrome for leather at lower pH. Therefore, aluminum salts can be incorporated into a chrome tannage to give a slight pre-tannage in the early stages; it is claimed that this results in a finer break. Aluminum will react with hide protein to produce a tanned leather. The bond between aluminum and hide protein is not as strong as the chromium-hide protein bond, and the stabilization of the hide protein or the tannage by aluminum is not sufficient, under normal circumstances, to produce a leather of a full-boil shrinkage temperature. Aluminum differs from chromium in that the basicity of aluminum goes from zero basic to 100% basic over a relatively narrow pH range. The addition of the salts of the hydroxy acids such as sodium tartrate or sodium citrate to the aluminum tannage greatly stabilizes the aluminum complex, permits tannage over a wider pH range, and will result in a much more stable tannage. Aluminum salts have the advantage of being colorless and are used in the production of furs and some white leathers. Formaldehyde is very often used as a supplementary tannage.

Alum tanning has been given renewed attention in recent years. There is a strong desire to decrease the concentration of spent chromium in the tannery discharge. Tannages with alum can be much more stable than previously believed. However, there has been little acceptance of alum tannage as a total replacement for chromium.

Figure 8 Shaving machine. The purpose of the shaving machine is to bring the leather to uniform thickness in a precise manner and with greater accuracy than can be obtained by splitting. The cutting blades are of steel and are kept sharp by a grinder in much the same manner as with a fleshing machine. The spiral blades on the cylinder aid in spreading out the skins. The leather is controlled by the operator. The feed roll is brought up against the cutting cylinder and is held away from the cylinder a predetermined distance by an accurate setting. During operation of the shaving machine the moisture of the leather must be such that the flexibility of the leather is maintained. If the leather is too soft, it will be dragged and torn by the shaving machine cylinder; if it is too hard, it will not be pliable enough to lie snugly against the guiding bolster. In the photograph the safety guard has been lifted on the machine to show the position of the leather bolster and the cutting cylinder. (*Courtesy Hebb Leather Co.*)

Zirconium tanning

Zirconium forms a 50% basic salt at very low pH (pH 2.0). Zirconium can be used as a tanning agent in itself and can produce a full-tanned all white leather that will stand a boil. This is usually accomplished by the use of sodium citrate or citric acid as a masking agent. The high acidity and high cost of the zirconium salt has limited its widespread commercial acceptance as a direct competitor to chrome tanning. Zirconium salts have recently been used extensively as re-tanning agents on chrome leather to produce desirable effects of fullness and tight grain.

LOOKING AHEAD

In the mineral tannage field it is evident that stabilization of the protein by mineral tanning agents results in leathers of greater chemical and thermal resistance than are presently possible from other tanning systems. The low cost of the mineral tanning agents, therefore, would indicate also that mineral tannages will become more widely accepted. It would be desirable to obtain a chrome tannage that is less sensitive to the conditions of pH and

Figure 9 Fleshing of chrome tanned sheepskins on a wheel. The use of an abrasive wheel permits accurate fleshing of sheepskins without danger of tearing. The operator can flesh each section of each skin as desired. (*Courtesy L. H. Hamel Leather Co.*)

temperature than the present systems. This could be accomplished if chrome tannage could be conducted at pH values close to the isoelectric point. The trends observed in chrome tanning have been towards higher pH values in tannages, higher concentrations of the liquor, and larger equipment for production efficiency. Chrome tannage or mineral tannage should continue to be a batch operation, but should employ large equipment and a high degree of automatic control.

In recent years, the leather industry has been dominated by its concern for pollution control. Research done in chrome tanning has concentrated much effort on the control of the fixation of chromium, recycling of chrome liquors, and the elimination of chromium in the discharge by end-of-the-pipe treatment systems.

The economic trends, the push toward the blue side concept, and the employment of much larger equipment for greater production efficiency have brought about an evolution in chrome tanning. Systems in which the unhairing, bate, pickle, and tan continue in the same container are common. The drastic chemical means that are used for these high speed systems would have been unthinkable before proper chemical control was available. Consistent quality is dependent upon balanced processes, consistent hide quality, and in many cases, completely automated control systems. There may or may not be additional significant benefits to be gained in higher speed systems, larger equipment, or slightly more efficient use of raw materials.

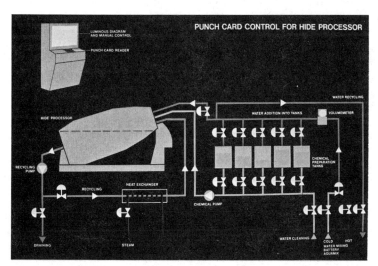

Figure 10   The Huni Process Control System for Hide Processors. Hide processors are particularly adaptable for full automation. The system shown can be used for the automation of automatic unhairing. Other tannery processes, particularly bate, pickle and tan, or retanning can be automated with similar systems. A programable computer permits process changes as needed. (*Courtesy of Hüni Company, Switzerland*)

The quality of leather may be more broadly altered by the use of relatively small amounts of chromium and the addition of other retanning agents. The cost of leather relative to other consumer goods is gradually becoming higher. This trend should force on the tanner an increased emphasis on leather quality.

REFERENCES

"Chrome Chemical Bulletin," Painesville, Ohio, Diamond Alkali Company, 1949.

Gustavson, K. H., "The Chemistry and Reactivity of Collagen," New York, Academic Press, 1956.

Gustavson, K.H., "The Chemistry of The Tanning Process," New York, Academic Press, 1956.

"Hide and Leather and Shoes Encyclopedia of the Shoe and Leather Industry," Chicago, Hide and Leather Publishing Co., 1941.

J. Am. Leather Chemists' Assoc., Supplement No. 5, Practical Aspects of Chrome Tanning (1957).

J. Am. Leather Chemists' Assoc., Supplement No. 6, Report of The Panel Discussion on Leather Making (1959).

"Leather Chemicals," Philadelphia, Rohm & Haas Company, 1955.

McLaughlin, G. D., and Theis, E. R., "The Chemistry of Leather Manufacture," ACS Monograph 101, New York, Reinhold, 1965.

Rodgers, A., "Practical Tanning," New York, Baird, 1922.

Somerville, I. C., "The Chemistry and Technology of Leather" (O'Flaherty, Roddy, and Lollar, ed.), Vol. 2, New York, Reinhold, 1958.

Stiasny, E., "Gerbereichemie," Steinkopff, 1931.

Suttleworth, S. G., "The Chemistry and Technology of Leather" (O'Flaherty, Roddy, and Lollar, ed.), Vol. 2, New York, Reinhold, 1958.

Thorstensen, E. B., "The Chemistry and Technology of Leather" (O'Flaherty, Roddy, and Lollar, ed.), Vol. 2, New York, Reinhold, 1958.

Thorstensen, T. C., "The Chemistry and Technology of Leather" (O'Flaherty, Roddy, and Lollar, ed.), Vol. 2, New York, Reinhold, 1958.

Thorstensen, T. C., "Symposium on Recent Advances in Mineral Tannages," Central Leather Research Institute, 1964.

Udy, M. J., "Chromium," ACS Monograph 132, Vol. 1, New York, Reinhold, 1956.

Wilson, J. A., "The Chemistry of Leather Manufacture," Vol. 2, New York, Reinhold, 1929.

Wilson, J. A., and Merrill, H. B., "Analysis of Leather," New York, McGraw-Hill, 1931.

VEGETABLE TANNING

9

The origins of vegetable tanning are lost in prehistoric times. Primitive peoples in all parts of the globe and from all the ages of the past have developed vegetable tanning systems based on materials available locally. If a raw skin was placed in contact with certain plants (bark, wood or leaves) moistened with water, it was found that the skin became stained and that the stained portions were protected against putrefaction. The broad availability of plants capable of producing this reaction and its simplicity of application led to the early development of a leather industry based on vegetable tannins.

Vegetable tanning and vegetable tanned leathers have been standards of leather production up until the fairly recent developments of the chrome tanning industry. Certain types of leather are made by the vegetable tan process because of the particularly desirable characteristics that the vegetable tannins produce. The leathers made by a full vegetable tannage are used for soles, belting, saddlery, upholstery, lining, and luggage. In addition to these, handicraft and some fancy leathers are also produced by vegetable tannage. The re-tannage of chrome tanned leather with vegetable tanning extracts is so common that it is found in most upper leather and also in garment leather formulations.

Vegetable tanning results not only in the preservation of the hide fiber, but also in a building into the fiber of certain characteristics of fullness of feel and resiliency which are characteristic of the tanning materials and methods used. In sole leather, the vegetable tannins impart not only body, fullness, and physical weight to the sole, but also molding characteristics so that a shoe will adapt itself to the individual foot. In upholstery, the vegetable tannins are used because the leather will have less stretch than if it were chrome tanned, and also, it will have a desirable soft, mellow feel. In lining leather, vegetable tannage gives the leather a hydrophilic character which will aid in the removal of perspiration from the foot.

THE VEGETABLE TANNINS

The vegetable tannins come from a wide variety of plants, and may be found in wood, leaves, nuts, twigs, and bark. The extract of a particular plant consists of a broad range of substances, and there is no such thing as a single tannin from a particular plant source. The material extracted from the wood not only contains many different tannins, but also starches, gums, and other materials. The extract is not a true solution but will contain suspended insoluble materials. These non-tannin materials also contribute to the leather-producing properties of the extracts. The structure of the vegetable tannins and the estimation of the tannin content of extracts have been a major field of work by leather chemists. The practical application of vegetable tanning has been based primarily on empirical skill.

Condensed tannins

The condensed tannins (catechol tannins) will increase in molecular size when boiled in an acid solution. Upon addition of alkali, the condensed tannins will be dispersed, are easily oxidized, and will develop red colors. Upon addition of iron salts, the catechol tannins will develop green colors. The tannins of quebracho, hemlock, wattle, and cutch are condensed tannins.

Hydrolyzable tannins

The hydrolyzable tannins (pyrogallol tannins) will disperse when boiled in an acid solution. Upon the addition of alkali they are less likely to develop red colors than are the catechol tannins. They can be more easily purified to relatively simple, light-colored tannic acids. Upon the addition of iron salts, the pyrogallol tannins will develop blue colors. The tannins of chestnut and myrobalans are of the hydrolyzable type.

The chemistry of the vegetable tanning materials is very complicated, and numerous methods have been used for the separation and isolation of the individual components of the vegetable tanning extracts. Attempts to make these separations have yielded more than fifty separate components of a particular extract.

SOURCES OF VEGETABLE TANNING MATERIALS

In the search for practical sources of tanning materials, almost every common plant has been subjected to extraction and the extract studied as a possible tanning agent. Only a small number of these materials, however, have shown true value as commercial products in the vegetable tanning field. The commercial tannins used in the United States are primarily imported either as the extract or as the plant material for extraction, depending upon economic and technical considerations. The extraction of domestic materials in the United States is done only to a limited extent at a few tanneries.

Some vegetable tannins

Quebracho. The name of this plant is *Quebrachia lorentzii.* The practical value of the tree for tanning material was discovered about 1870. The tree grows in South America, predominantly in Argentina. It is a very slow-growing tree, taking approximately one hundred years to reach maturity. It grows to an average height of approximately 40 feet but may grow as tall as 60–75 feet. The wood is very hard and dense, and it is said that the name of the tree comes from the words "quiebra hacha," meaning "break axe," from the characteristics of the hard wood. The top of the tree has a peculiar

TABLE 1

Analysis Data of the Most Important Tanning Materials Used Today in the Air-Dried Condition

(Water Content = 12–14%)

	% Tans	% Soluble Non-Tans			Percentage of Tans in Total Solubles
		Sugary Matter	Non-sugary Matter	Total	
Oak bark	10	2.6	2.9	5.5	65
Pine bark	11.5	5	2	7	62
Mimosa bark	36	4	3.5	7.5	83
Mangrove bark (east)	36	0.8	8.2	9	80
Mangrove bark (west)	24	2	13	15	61
Valonia	29	3	7	10	74
Valonia cups	40	3.5	9	12.5	76
Myrobalans	34	5.5	8.5	14	71
Myrobalans, cored Fruit pulp	50	8	9.5	17.5	74
Divi divi	41.5	10	9.5	19.5	68
Algarobilla	43	8.5	13	21.5	67
Knoppern	30	0.6	6.9	7.5	80
Sicilian sumac	26	4.5	9.5	14	65

TABLE 1 (cont.)

Analysis Data of the Most Important Solid Extracts

	Tans	Non-Tans %	In-solubles %	Water %	Tans in Total Solubles %	Average Ash Content	Average pH Value
Quebracho ordinary	65	5	8	22	93	1	5
Quebracho sulphited	70	8	0	22	90	5.5	6
Oakwood extract	61	22	2	15	74	2.7	3.7
Chestnut wood extract	66	16	2	16	81	1.5	3.5
Mimosa extract	63	16	1	20	80	3	5.2
Mangrove extract	59	17	1	23	78	5	5.4
Valonia extract	69	21	1	9	77	5	3.8
Myrobalans extract	59	18	5	18	77	3	3.2
Sumac extract	64	29	1	6	69	5	4
Block gambier	51	14	5	30	78	3.5	4.3
Cube gambier	55	17	12	16	76	4.5	4.7
Pine bark extract	54	34	2.5	9.5	61	2.5	4.3
Redunca (Myrtan)	60	16	approx. 12	20	79	4 – 5	4

U-shaped crown. It is a fir evergreen type and is somewhat scattered. Usually, four or five trees are found to an acre. The heartwood at the center of the tree is used for the production of quebracho extract, which yields only 6–8%.

The extract as received has a tannin content of about 63–70% in good quality extracts; the normal material as extracted has insoluble matter to the extent of 6–10%. During the extraction the heat results in further condensation of some of the extracts (since quebracho is a pyrogallol or condensed tannin the insolubles will be increased). In order to decrease the insolubles in quebracho and improve its color and ease of application, much of it is clarified by treatment with bisulfite prior to use in leather production. The reaction of the bisulfite decreases the insolubles to 1% or less, and the non-tans are increased. Bisulfiting of the extract causes a change in the tanning characteristics in that the astringency of the tannin is decreased.

Quebracho extract, having a very low percentage of non-tannins, has a strong tendency to fix with the hide, and thus it is an astringent tannin. Solubilization of the extract by bisulfiting decreases this tendency, improves the color, has a tendency to penetrate the hide much more quickly, but gives a softer, more empty leather. The leather has a tendency to redden on exposure to light.

Quebracho is by far the most important and the most widely used extract in the American tanning industry. In 1966 the quantity of quebracho extract imported into the United States was approximately 80 million pounds, and in addition to this, over 12 million pounds of quebracho wood was also imported. Quebracho extract has a natural pH somewhat greater than 6 and an average molecular weight of 2400. Sulfited quebracho also has a natural pH value in excess of 6 and an average molecular weight between 700 and 800.

Quebracho extract is used in the manufacture of heavy leathers and in the vegetable re-tanning of chrome tanned upper leathers. It is also widely used as the main tanning agent for all forms of vegetable tanning, and its characteristics are modified by blending with other vegetable tanning extracts.

HISTORY OF THE DEVELOPMENT OF THE QUEBRACHO EXTRACT INDUSTRY

The discovery of quebracho as a tanning agent is attributed to a German botanist who was hunting orchids in the forest jungles of Argentina. He discovered a brook which ran a rather red color, which was strange for that part of the world. Following the stream he came to a sawmill where railway ties were being made. Rain falling on the red sawdust imparted a red color to the water. The sawdust was analyzed and was found to contain approximately 25% tannin. The dense wood was cut into railroad ties and was so hard that the spikes could not be driven into the ties; holes had to be drilled into the wood to take the spikes. The corrosion of the steel spikes by the tannin of the wood caused the spikes to be held firmly in place. It was said that such ties would last forty to fifty years.

By the early 1870's the French had a commercial extract plant for the extraction of quebracho wood. The first quebracho logs were imported to the United States in 1897. This fifty-ton shipment was given to a dyewood company. The company had a very difficult time handling the hard wood, and they decided finally to reduce the logs to a size for the chippers by breaking them apart with dynamite. The brittle wood shattered under the explosion and was eventually processed into a heavy extract containing about 36% tannin. The tanners of the Newark area immediately recognized the tanning extract as being revolutionary, and within fifteen years the methods of tanning in the United States were revolutionized. The growth and acceptance by the tanning industry was so rapid that in 1904 a single order was accepted for 1000 carloads of the liquid extract. This order involved the cutting, shipping, and extracting of 30,000 tons of quebracho wood.

Wattle extract (Mimosa). The second most important extract, from the point of view of total quantities of material used in the United States, is wattle extract. Wattle is a catechol tannin coming from the bark of the trees of the various acacia species. In 1966, the total quantity of wattle extract imported was approximately 25 million tons and the bark for extraction was an additional 13 million tons. The extract was imported from the Republic of South Africa, with a small quantity coming from Brazil. The plant was originally found in Australia, but it is grown commercially on a large scale in South Africa. The main source of wattle extract is the *Acacia mollissima,* or Black Wattle. It will analyze somewhat over 60% tannins and 18–20% non-tannins. The extract contains some sugars and non-tannins and is usually low in insolubles. Penetration is quite good, and it is often used in the re-tannage of upper leathers, as well as in the production of heavy leathers. It has a natural pK value in excess of 6 and has an average molecular weight between 1600 and 1700. The color of the extract is somewhat less red than quebracho extract.

Myrobalans is obtained from the fruits of the *Terminalia chebula.* It is a hydrolyzable tannin with a natural pK value of 4.5 and a molecular weight of 1900. The fruits are dried nuts, rich in tannin, and also contain highly fermentable sugars. The tannin content of the nuts is approximately 30–40%. The tannin has a deep yellow color and gives a greenish cast to the leather. The popularity of myrobalans in the sole leather industry is attributed to the characteristic of the material to ferment and produce acids. The low pH of the liquor, as well as acids produced by fermentation, is helpful in obtaining good fixation of the vegetable tanning materials through the control of the pH values of the tan liquors. Myrobalans is a pyrogallol tannin.

Sumac. Sumac is the extract of the leaves of *Rhus coriaria.* This is a very light-colored extract and is a very mild tanning agent. Because of its light color, it is used primarily for re-tanning and mordanting. It is a very weak pyrogallol-type tannin, and the resulting leather is quite stable to acids. The plant contains approximately 25% tannin.

Gallnuts. Gallnuts are formed by the swelling of the leaves and buds of Turkish oak trees when punctured by certain insects. These parasitic growths contain approximately 50–60% tannin. They are a very pure, light-colored tannin and are the source of the tannic acid of commerce. Approximately one million pounds of gallnuts were imported to the United States during 1966. The extract is used as a mordant and also in the manufacture of inks.

Gambier. Gambier is one of the mildest and mellowest tanning agents known. It comes from the leaves and twigs of the bush Nauclea gambir found in the East Indies. The trees grow to a height of approximately ten feet, and the leaves and twigs are extracted for tannin. The yield of tannin from the leaves is too low to warrant the shipping of the leaves and twigs, so the tannin must be extracted in the country of origin. It grows only within a narrow belt a few degrees from the equator. Gambier started to gain popularity just prior to the beginning of World War II, and plantations were established in the islands of the East Indies. During the war the market for gambier was lost with the occupation of the area, and the establishment of the gambier plantations has never been rebuilt. The plant grows very quickly, and with the excellent quality tannin obtainable from the gambier, it could be developed as a source of vegetable tanning material if economic conditions would warrant it.

Mangrove. Mangrove extract is obtained from the bark of various species of the mangrove tree of the species rhizophora. The material is also known to tanners as "cutch." The solid extract contains approximately 55% tannin. The product is obtained in Borneo, Kenya, Mozambique, and India. The tannin content of the plant varies between 8 and 30%, depending upon the species. The tannin is of the catechol class and results in a somewhat dark red color. It has a hard, brittle tannage and is used in blends with quebracho and other materials to increase plumpness and yield in the tannage of heavy leathers.

Hemlock. Hemlock is an evergreen tree indigenous to North America, and the bark has been used as a vegetable tanning material, particularly in Michigan, Minnesota, and Wisconsin. It is a catechol tannin, giving a firm red leather. Its use is very rapidly being discontinued and very soon it will probably be of only historic interest.

Chesnut. Chestnut wood was the most important domestic source of tannin in the United States and was the reason for the establishment of the sole leather industry in Virginia, North Carolina, and Tennessee. The availability of chestnut from domestic sources has been decreasing, due first to the blight that has hit the chestnut tree, and second to the increased costs of obtaining the wood. The chestnut tree, *Castanea dentata,* has all but died out in the United States. As a result, American tanners have turned to the imported chestnut extracts, and in 1966 approximately 17 million pounds were imported from France and Italy. The tannin is hydrolyzable and has a natural pH value of about 5 and a molecular weight of about 1500. The extract tends to give a deep yellow color. It is used in the manufacture of heavy leathers.

Valonia. Valonia is obtained from the Mediterranean oak, *Quercus aegilops,* which grows in the eastern Mediterranean area of Greece, Turkey, and Israel. The cups and beards appear very much like the American oak acorn and produce a pyrogallol tannin with a yield of about 30% from the cups and about 40% from the beards. The extract possesses some filling properties and gives a solid, firm leather with a pleasing color. It is used in blends in the manufacture of heavy leathers.

Lignosulfonates. In the manufacture of paper, lignin of the logs of the wood chips is released by cooking the chips with bisulfite in accordance with the Kraft process. This results in a reaction of the bisulfite with the lignin to form a lignosulfonate material and a breakdown of the wood to form cellulose fibers and the lignin liquor. The lignin liquor, containing sugars, is available in a very large supply and is a problem in stream pollution. The close relationship between the chemical structures of lignin and the vegetable tannins indicates the possibility of this material for use as a leather-making agent. However, the phenolic groups are esterified in lignin, and instead of OH, there is predominantly OCH_3 ether. In the preparation of lignosulfonates for tanning, the sugars are usually fermented from the solution and the lignosulfonate recovered as the salt. The material can be made available in the dry form and has a strong affinity for leather. It does not produce true leathering effects, but when absorbed by the skin, it will have a filling action. Lignosulfonates are commonly referred to by the tanners as spruce extract. This is not to be confused with the extract of spruce bark, which is not a significant source of commercial tannin.

(a)   (b)

Figure 1   Vegetable tanning materials. The vegetable tanning materials are shown in the condition ready for extraction. The photos are taken the same way in each case, so the sizes are comparable. (a) Tara pods. (b) Wattle bark. (c) Logwood chips. (d) Gallnuts. (e) Sicilian sumac. (f) Myrobalans.

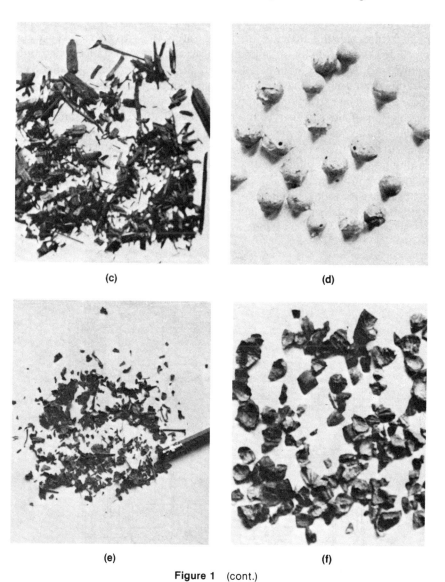

(c)

(d)

(e)

(f)

Figure 1   (cont.)

EXTRACTION OF VEGETABLE TANNING MATERIALS

The availability of vegetable tanning materials was of prime consideration in planning the location of a tannery for the heavy leathers of a century ago. The tanners extracted the tannins from the local barks and used the extracts directly in their leather production. The problem of fermentation of the tanning extracts and the difficulty of concentrating the extracts to save weight precluded the possibility of shipping extracts for great distances. Chestnut tree extract was very common through the areas of Pennsylvania and Virginia, resulting in the establishment of large sole leather factories in these areas. The extraction from the bark was conducted in open tanks by percolation of hot water over the bark. Several successive washings were made with the weak extract solution from previous runs. By this method the concentration of the tanning materials could be increased. The tan liquors resulting were rather dilute, and this was a significant factor in the design of the tanning process.

For efficiency today, it is necessary to extract the tanning material in a centralized factory, either in the United States or at the country of origin, and the extract should be made as strong as possible to save weight in shipping. A

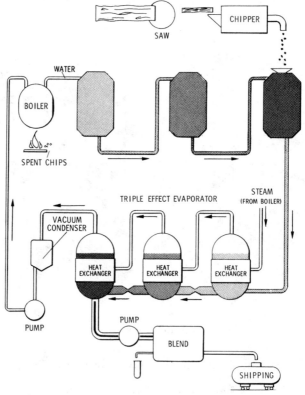

Figure 2 Flow diagram of an extract plant. Vegetable tanning materials are extracted by the counter-current system. This yields as strong liquors as possible, yet the spent materials are well extracted. The extracts are concentrated by evaporation. The triple effect evaporator conserves steam and prevents overheating of the extract.

modern extract plant, therefore, operates with a battery of extractors working on a counter current system. The fresh water, under close temperature control, enters into the first extractor and comes in contact with tanbark that has already been passed over by six or eight volumes of water. As the water passes over successive layers of vegetable tanning material, the solution becomes stronger and stronger and is in contact with fresher and fresher materials. The last material having contact with the tan solution is the fresh bark entering into the last extractor. Through a manifold system, the extractors are continually being loaded and discharged and the cycle is run without interruption from one extractor to the next. The spent tanbark is burned in a boiler especially designed for the purpose, and the steam generated is used to heat the hot water for the extraction process and for the evaporators.

The extract, after leaving the extractors, is fed into multiple-effect evaporators which are designed in such a way that the steam generated from the boiling water in the first evaporator is condensed in the second to heat the extract. The second effect of the evaporator then generates more steam which is condensed in the third effect, further evaporating more water from the extract. Each successive stage of the evaporation operates at lower temperatures as the vacuum inside the evaporator is increased. The heat balance of the system automatically provides for the temperatures and pressures in the various stages of the evaporator. In the last effect of the evaporator the boiling temperatures are significantly lower than in the first stage, and under this high-vacuum system the extracts can be concentrated to a very high solids content without burning the extract. The condensate from the evaporators is reheated and re-used in the process. By this means a minimum quantity of fresh water can be introduced and the mineral content of the extract is kept at a minimum. The extract from the evaporator is analyzed for tannin content and is then blended with other extracts to make the standard grades.

Spray-drying

Spray-drying is a common means of preparing vegetable tanning extract. The method is also used in preparing foods, such as coffee, cocoa, and many other materials. In spray-drying, the liquid is raised to a relatively high temperature, and it is as concentrated a solution as possible. It is then sprayed into a large chamber where the warm liquid falls through the air and the moisture is evaporated from the surface. The result is very light spheres of extract in a relatively uniform size. Spray-dried extracts are easily soluble and can be conveniently added in a drum system as solid material. Spray-dried extracts are more expensive than those extracted by the triple effect evaporated technique, since the economy of energy is less.

CHEMISTRY OF THE VEGETABLE EXTRACT SYSTEM

The vegetable tanning solutions and their behavior arc very complicated from both a chemical and a physical point of view. In considering the mechanisms and the behavior of practical tanning solutions, the colloidal phenomena, as well as the better-understood chemical reactions, must be taken into consideration. The three most important factors affecting a tanning operation are pH, concentration, and temperature. All thse factors will affect both the colloidal properties of the tanning medium and the chemical reaction of the tannin molecules.

Effect of pH

The weak acid nature of the tannin molecules will aid in the increase in charge on the particles with increases in pH, and the colloidal particles will disperse as the weak vegetable tannin compounds become more soluble. The pH of the solution will also affect the color of the extract. At higher pH the greater solubility will generally result in a darker color and less cloudiness than at lower pH.

Effect of concentration

The higher the concentration, the greater the reaction or fixation of the vegetable tanning material by the hide protein. In practical application, high concentrations of vegetable tanning materials may result in a very large particle size of tannins which will cause precipitation (excessive deposition on the surface of the leather) and thus block the penetration of extract into the inside of the skin.

Effect of temperature

The higher the temperature, the greater the dispersion of vegetable tanning materials and the greater the rate of the reaction of the vegetable tannin materials with the hide protein. One factor, the decrease in particle size, will aid in the penetration of the vegetable materials into heavy leathers. On the other hand, the increased rate of fixation may result in excessive deposition on the surface of the skin. Practical tanning by the vegetable processes will depend not only on the nature of the vegetable tanning material and the condition of the hide, but also on a skillful balance of pH, temperature, and concentration.

Significance of the chemical analysis of the vegetable tanning materials

One of the original purposes of the formation of the ALCA* was to establish standard methods of determining tannin content of vegetable tanning materials. Since the materials are bought by the pound, and since hides are purchased and sole leather sold by the pound, it is important that the tannin content, or the substance available in the vegetable extract for fixation on the hide, be known. The tannin analysis method, as set up by the ALCA, is similar to other vegetable tannin analysis procedures used for the market evaluation of an extract in that it is based on the absorption of materials from the extract by hide protein. It is not based on a chemical analysis of a true tannin molecule.

In the ALCA tannin analysis method the extract is made up to a solution containing what is anticipated to be approximately four grams of tannin per liter. A portion of this material is evaporated to dryness to give total solids. After a carefully standardized filtration technique, another portion of this material is evaporated to dryness to give the soluble solids. A third portion of the vegetable tannin solution is shaken for a given period of time with a special preparation of purified, powdered, hide under closely controlled conditions. At the end of this time the solution is separated from the hide powder, filtered, and a portion evaporated to dryness. These three analyses, with proper corrections for volume changes, give the total solids, soluble solids, and the non-tannins. The subtraction of the non-tannins from the soluble solids gives, by definition, the tannin analysis. The results are reported as percent by weight. In addition, the tannin may be expressed as a percentage of the soluble solids. This is known as the "purity" of the extract. Vegetable tannin materials, bark leaves, etc. are analyzed by this same method after a specified extraction procedure.

The method gives consistent results, and for a particular extract they are a true measure of its active component content. The tannin analysis method, however, does have its drawbacks. It is not a measure of leather-making ability and does not describe the characteristics of the extract. Caution should be used, therefore, in the application of a tannin analysis to a practical problem. It should be recognized that this analysis should be used only to compare two samples of the same type of extract.

Two classes of material often subjected to tannin analysis give misleading results. These are the exchange syntans and the lignosulfonates. The exchange syntans may have far greater leather-making power, as judged by

*American Leather Chemists Association

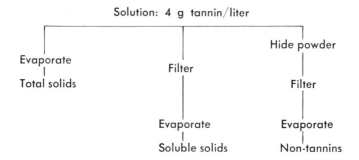

Figure 3 Flow diagram for tannin analysis by ALCA method. The analysis of tannin is based on the absorption of the active components by a specially prepared hide powder. Differentiation is also made for the truly soluble materials and those held in colloidal suspension. The method is designated for vegetable tannins and should not be applied to syntans or other similar materials.

changes in the fiber character, than a "tannin analysis" would indicate. The lignosulfonates have very little leather-making ability as judged by fiber condition and yet may show a high "tannin" content.

A further point of confusion on the same subject is the relationship of the dyewoods to the vegetable tannins. Dyewoods are manufactured in a manner similar to vegetable tannins and are closely related in that they are flavins and are vegetable extracts with a large quantity of phenolic hydroxyl groups. They differ from vegetable tannins in that they have a predominance of highly colored components and they contribute bright colors—blues, reds, and blacks—when applied to leather. They would show tanning value when subjected to tannin analysis, and it has been stated that they do have tanning properties. Tanners attempting to make leather with dyewood extracts alone, however, would be greatly disappointed in the leather quality because, like the lignosulfonic acids, their true leather-making properties are very limited.

THE THEORY OF VEGETABLE TANNING

The mechanism of vegetable tanning is one that has been studied by leather chemists for decades, yet this work has not yielded clear theories of vegetable tannin fixation which are of direct value to the practical tanner. First, the vegetable tannins themselves are such complicated mixtures of components that the materials cannot be defined by relatively simple structures. The second factor is the colloidal behavior of the vegetable tannin materials. A third complicating factor is the hide protein. The condition of the protein, as a result of previous treatment, affects not only its physical structure but also its chemical reactivity. The degree of opening of the fibers, the availability of reactive groups for vegetable tanning, the degree of hydration, the presence of salts, and the extent of swelling of the skin are all-important in practical vegetable tanning. The complexity of the phenomena involved, and also the further complication of working against one another, have left the problem of the mechanism of vegetable tannage and the relationship of possible mechanisms to practical tannage in a state of near-chaos. In spite of the tremendous amount of work that has been done, there is still disagreement among the experts as to what groups of the protein are involved in vegetable tannage and even, in some cases, whether vegetable tannage is a chemical action or strictly a physical absorption. It would probably be right to say that both chemical and physical phenomena are involved in the reactions; it is the combination of these factors and the relationship of the various vegetable tannins to both phenomena which determine their practical application.

The generally accepted mechanism for vegetable tannage is through hydrogen bonding to the CO—NH linkage of the protein through the phenolic hydroxyl group of the vegetable tannin. Many other side reactions have been postulated, but this general reaction has been considered to be predominant by most workers. The curve of the fixation of vegetable tannin materials as a function of pH indicates a slope downward from the strong acid range to a minimum near the isoelectric point, a rise to a minor peak, and a fall. This is the initial tannin fixation for most vegetable tanning materials with hide protein. There are several factors involved which would explain this characteristic curve. The availability of hydrogen bonds or hydrogen atoms on the protein and on the vegetable tanning material is of prime importance. pH values for the vegetable tannins are from 5–7,

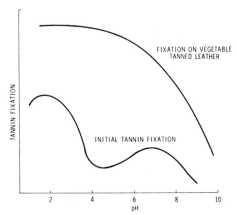

Figure 4   Tannin fixation is related to pH. During the initial stage of tanning the rate of fixation vs pH gives an S-shaped curve as indicated. This deviation from the acid-binding curve of the protein is due to replacement of water from the protein. Later, the acid-binding effect becomes dominant and the upper curve results.

depending on the vegetable tanning material, as indicated in Table 2. Thus, the vegetable tannins are essentially un-ionized through the entire vegetable tanning range, and the ionization of the phenolic hydrogen is not the main factor for fixation. The protein, on the other hand, increases its hydrogen ion fixation with the lowering of the pH; therefore, fixation of the vegetable tannin is dominated by the reaction of the hydrogen ion with the protein, rather than the hydrogen ion on the vegetable tannin material. At the high pH values, in the neighborhood of pH 5, the hydroxyl groups on the vegetable tannin material become ionized and the fixation of the vegetable tannin decreases distinctly.

The S-shaped curve near the isoelectric point emphasizes the fact that the reaction is not strictly chemical. The swelling characteristics of the protein are such that hydration of the molecule is least near the isoelectric point. Increase in pH causes an increase in hydration and in the fixation of the vegetable tanning material at pH values just above the isoelectric point, thus introducing a bump in the curve. As vegetable tannage proceeds, the fiber is no longer dominated by water of hydration, but rather has changed in character as a result of the fixation of the vegetable tannins. The hydration factor becomes less significant, and the eventual fixation curve is a smoother sweep from the neutral pH range to a high fixation at the strong acid range.

TABLE 2
Vegetable Tannin pH Values as per Gustavson.

Material	Type	Purity	pK	Molecular weight
Myrobalans	Hydrolyzable	86	4.5	1900
Chestnut	Hydrolyzable	84	5	1550
Gambier	Condensed	87	5	520
Wattle (mimosa)	Condensed	95	>6	1600–1700
Quebracho	Condensed	95	>6	2400
Sulfited quebracho	Condensed		>6	700–800

Stoichiometric considerations

The quantity of vegetable tanning material necessary to bring about proper tannage is much greater than one would expect from a purely chemical point of view. The vegetable tanning material molecule is very large, and in spite of its possibility of multipoint attachment, it cannot reach the available sites in the protein. The large size also prevents fixation of a second vegetable tannin molecule to an available site. Vegetable tannage can be looked upon as being a partial reaction between the protein and the tannin, the reaction being blocked as a result of the physical size of the hide fiber and of the vegetable extract particles. As tannage proceeds, additional vegetable tanning materials are deposited on and between the hide fibers, resulting in eventual coating of the fibers and filling of the voids of the hide with vegetable tanning materials. The tannage becomes effective (1) by the chemical reaction between the vegetable tanning material and the hide protein, (2) by coating of the fibers and isolation of the reactive groups, and (3) by filling of the physical voids in the leather.

PRACTICAL VEGETABLE TANNING

The techniques employed by tanners to produce different leathers are based on controlled penetration of the vegetable tanning materials into the hide or skin. The characteristics of the various vegetable tanning materials with respect to their ease of dispersion, degree of penetration, and degree of filling properties are related more to their colloidal character than to the chemical nature of the tannins involved. In the vegetable tannage of light leather, particularly in a drum, the chemical aspect is most important. Here the tanner is trying to obtain a full, soft tannage with a minimum amount of material. In heavy leather, where filling action and weight increases are important and a rocker system is employed, the colloidal character of the system is dominant. This explanation of the mechanism of tannage emphasizes the deficiencies of the vegetable tannin analysis as a measure of the true tanning characteristics of a material.

Vegetable tanning of sole leather

In the manufacture of sole leather, vegetable tanning is the predominant process for two reasons. The first is that it imparts to sole leather the property of moldability. When made into a shoe and worn, the leather will take up some moisture from the wearer's foot and will gradually yield to pressures and conform to the foot. This property of sole leather is one of the main reasons for the comfort of leather shoes, and it is chiefly because of this that vegetable tannage is used in its manufacture. The second function of vegetable tanning in sole leather manufacture is the buildup of solid materials inside the hide to give more physical weight and greater wearing characteristics. The tendency of the sole leather tanner to add weight and substance to his leather (since the leather is sold by the pound) has been, in the opinion of many, a deterrent to progress in sole leather manufacture.

In the vegetable tannage of sole leather, very heavy hides are used. These hides are given a long liming period to open up the structure and make them more suitable to receive the vegetable tanning components. The bating and deliming process brings them to the slightly acidic pH range, and they are then introduced into the vegetable tannage rocker system. It is the usual practice to segment the hides at this point to remove the double shoulders or the bellies, or both, so that the greatest economic value can be achieved from various parts of the hide. The initial vegetable tanning liquors are quite dilute and consist of the spent liquors from previous tannages. Normal vegetable tannage by the traditional method is in a rocker system, where the hides are hung on racks in pits containing the vegetable tanning materials. A gentle rocking action keeps the hides in constant motion in the tanning solutions without folding or flexing. The vegetable tanning process continues for approximately three weeks, during which time the leather is placed in solutions of gradually increased strength. This is done by shifting the hides from pit to pit of increasing concentration of tanning materials, or by pumping the solution from one pit to another. The vegetable tanning process, as employed here, is a counter-current take-up of vegetable tannins, as opposed to a counter-current extraction in the modern production of the vegetable tanning extracts described earlier in the chapter. At the end of the tanning time the tannin is thoroughly struck through the skin and the skin is tanned.

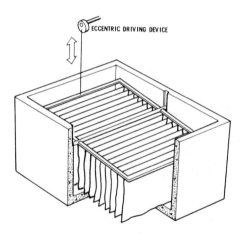

Figure 5 Plan of a single vegetable tanning rocker. Vegetable tanning rockers are built in batteries of twenty or more. The pits contain extracts at different strengths and the hides are moved from one pit to another. In some systems the liquors are pumped. The hides are kept in constant motion by the rocking action of the eccentric. A large tannery may have several hundred rocker pits.

In further sole leather processing, more vegetable tanning materials are introduced by treating the skins with very concentrated liquors, thereby building up the concentration of the solids in the voids of the skin. The increase in fixation of the vegetable tannins near the end of the tanning period may be accomplished by lowering the pH. It would be desirable to start the initial tannage at a relatively high pH in order to allow good penetration of the dilute tan liquors. At the end of the tannage, penetration can be accomplished best by a high concentration of tanning materials.

Before the days of proper knowledge of pH control it was customary to allow the souring of the vegetable tannins to proceed through natural fermentation of the sugars (in the extracts) to acids. This controlled fermentation was very difficult to keep in proper adjustment, and quality control of this system was a matter of luck. The present system employs very close control of pH by the addition of disinfectants, proper balance of extracts, and added organic acids. With the changes in technology toward low labor cost and high-speed production, a number of systems have been developed based on close control of temperature, pH, and salt content of the tanning materials. This careful control permits rapid penetration of the vegetable tanning extracts. Also, the use of drums is becoming more common.

Syntans, of both the replacement and the naphthalene types, are used to aid in the dispersion of vegetable tanning materials and to facilitate penetration of the extracts into the skin. For the manufacture of heavy leathers, very fast drum tannages have been reported which promise to decrease the time of tanning from two to three weeks down to less than one day.

Sole leather has been a standard item for decades and there has been little real difference in quality from one manufacturer to another. However, customer (consumer) demands for light-weight sole leather with good performance characteristics are bringing about a change in attitude regarding the philosophy of sole leather quality.

Two major changes are occurring in the sole leather industry. First, the increase in pressures for high-speed production is forcing the use of more modern and efficient equipment, and second, changes are taking place in standards of quality. As a result of these two factors, there will probably be a major change in the nature of sole leather within the next few years. In the author's opinion, this is long overdue.

Vegetable tanning of upholstery leather

The desired characteristics of upholstery leather are quite different from those of sole leather. In upholstery leather, the soft breathability characteristics and limited stretch favor the vegetable tannages. There is no desire, in the vegetable tannage of upholstery leather, to build up the body of the leather or to fill the voids. The tannage may be conducted in rockers in the same manner as the sole leather system. However, concentrations and types of liquors used are different in that a fast penetration of the soft leather is desired. Exchange tannins can be blended with the vegetable tanning materials very effectively in upholstery leather manufacture; this results in the fast penetration and softer leather desired for this type of production.

After the initial vegetable tannage has been completed and the leather is tanned all the way through, it is split into several different layers: the grain, or top grain leather; the second layer, which is the deep buff; the third layer, called the slab; and the fourth layer, the split. These individual layers may then be vegetable re-tanned in a drum, colored, and fatliquored to the desired characteristics. Pressures for high-speed production in vegetable tannage have been felt by the upholstery leather tanners, and in recent years it is becoming increasingly evident that drum tannages are desired. In order to avoid excess drawing in the mechanical action of the drum tannage with vegetable extracts, a pre-tanning action with a small quantity of chrome tanning agent may be employed.

Vegetable tanning of light leathers

Numerous light leathers are subject to vegetable tannages for the purpose of developing special characteristics. These include sheepskins for lining leathers, hatbands and bookbinding leathers, calfskins for wallets and tooling leathers, reptile and fancy leathers for handbags and novelty items. The principles of operation for their tannages follow the same pattern as the drum tannage of upholstery leather. The preliminary tannage of the pickled pelt by chrome may be used, or tannage may begin with pre-tannage of a syntan of either the naphthalene or phenolic type. The vegetable tannage may be conducted with a blend of extracts to give the desired characteristics of feel and color to the leather. The choice of extracts is governed by the characteristics of the leather to be produced. The speed of penetration is of limited importance, and the tannage is controlled by control of the liquor ratio, temperature, and pH of the solution. The light leathers tanned for these purposes can be tanned in approximately a 24-hour cycle in the drum to produce the desired results.

Process trends in vegetable tanning

Two main concerns in vegetable tanning have caused changes in processing in recent years: inventory costs and pollution control. Vegetable tanning is a long process, requiring large inventory in process, which is a financial burden on the tanner. Financial pressures and equipment limitations have forced faster tannages. These tannages are based on a balance of conditions of pH concentration, temperature, and added materials. Pretanning is important in most modern tannages.

The three most important phenomena which affect the rate of tanning are the electroletic (ion) equilibrium, the rate of diffusion, and the fixation or tanning. These processes are interrelated.

The pH is one of the most important factors in the electroletic equilibrium. That is, there must be a removal of lime at the initial stages, while maintaining a slightly higher pH in the solution than is desired at the end of the tannage. The more modern processes, deliming, may take place by the addition of Calgon (a polymeric phosphate) or boric acid, ammonium chloride, or sodium bisulfate. The pelt may be delimed in a drum to bring about the desired pH and the removal of alkalinity from the center of the skin. The solution acidic conditions desired are: pH around 4½ to 6 with the pelt below pH 8 throughout the skin. Proper balance of concentration of liquors, the addition of syntans (Chapter 10), and the addition of acids for a proper pH adjustment are the basis of the modern tannages. Three specific tannages now in use are good examples of modern thinking.

Liritan tannage

This system is very popular in that it is a "no effluent" system. It is a fast tannage and relatively easy to control. Process is done in two or three stages. First is a deliming process, using an acid Calgon pickle. This is approximately 5% Calgon and the limed pelt is delimed at pH 2.8 to 3.0 by an addition of sulfuric acid. An open pit pickle is usually necessary to ensure pH near 3 and equilibrium. The spent pickle liquor is kept and refortified by the addition of approximately 2¼% Calgon on limed weight and about 1–1½% sulfuric acid. These quantities are finalized by the tanner to best suit his individual conditions.

Drained pickled pelt is transferred to a warm vegetable tan with a pH of between 3.2 and 3.5 and a density of approximately 1.1 or 100° barkometer. The temperature is kept at about 35° C. Tannage proceeds for approximately three to ten days. This tannage was originally developed for Mimosa, but since then has been adapted to a number of different tanning materials. The vegetable tanning liquor maintains its strength by refortifying with the addition of more vegetable tanning material. The solutions stabilize on each repetitive bath. The quantity of Calgon in the tanning bath will be maintained at slightly less than 1%. The pH will stabilize between 3.2 and 3.5 as desired.

In some cases, it has been found convenient to use three steps. Between the Calgon pickle bath and the warm vegetable tanning bath, a vegetable tanning step is added, using a coloring vegetable bath of about 20° barkometer. This step produces slightly better leather, and is easier to control. However, it will result in some effluent, since the intermediate stage may not be refortified and maintain consistent concentration in the low strength liquors.

Liritan System

(1) Deliming in pit
5% Calgon   H$_2$SO$_4$
pH   2.8–3.0
   Reinforce with Calgon and H$_2$SO$_4$ to keep constant conditions.

(2) Coloring bath (optional)   20° Bk

(3) Vegetable tan 3–10 days
pH 3.2–3.5
density   1.1   (100° Bk)
35° C
Stabilized to approximately 1% Calgon

Modern pit-drum tannage

A pit-drum type of vegetable tannage that is achieving wide-spread acceptance is the modern pit-drum process, which involves a pretannage in pits and completion of the tannage in drums. The process takes place in four steps. First, the pelt is delimed and lightly tanned in a drum with a syntan, the phenolic type, and tannage takes place in about one and one-half hours. About 2–3% salt is added to prevent excessive swelling and about 1% of the pretanning syntan is added. The pH is adjusted to about 5.

Following the pretannage, the hides are removed and put in the circulating drum system where the concentration is about 40° barkometer and pH is about 5, using chestnut, Mimosa, or bisulfited quebracho. A proper blend of extracts is necessary at this point to get good penetration and avoid drawing of the grain. Timing in the coloring pits is approximately between one and two days.

During the main tannage, the density of the solutions is brought up to about 70° barkometer and the pelts are in the pits for about eight to twelve days. The pH is dropped to about 4.5 during this period of time. A series of pits is used and restrengthening of the solutions maintains approximately constant conditions throughout the series. At the end of the pit tannage, the hides are placed into a drum and tanned at 100° barkometer pH 3.8 for about two to three days. A natural acidic tan liquor is desirable.

Modern Pit Drum Tannage   15 Days Total

(1) Delime   Pretannage   Drum
Phenolic Syntan   1½ hours
2 – 3% salt
pH 5

(2) Tannage   Circulating drum
40° Bk   pH 5
Mimosa, bisulfited quebracho, or chestnut 1 – 2 days

(3) Main tannage   Pits in series
pH   4.5
8 – 12 days
70° Bk to pH 3.8
100° Bk
temp. 37° C

(4) Drum tan
pH 3.8
100° Bk

Chrome pretannage

A third type of tannage is the chrome pretannage system. In this case, the pretannage chromium is a somewhat conventional chrome tannage. The hides are placed into the drum and delimed with ammonium sulfate and about 1½% sulfuric acid. The chrome tanning powder is then added (1% dry chrome tanning material (33% basic) or ¼% Cr_2O_3 based on the weight of the limed pelt). The tannage is run intermittently and the next day the excess chrome liquor is removed by draining off. The process can be continued in the same drum with a gradual addition of vegetable tans. In this case, the dry vegetable tanning powder is added, starting out with about 5% Mimosa powder, and after a half hour, an additional 10% is added in gradual increase in concentration of the vegetable tan powder. After about 40% of the dry powder has been added, penetration will usually be complete in approximately twenty-four hours. At this point, addition of some bisulfite or formic acid will aid in proving the color. At the end of the tannage, the adjustment of the pH to 3.2–3.3 is gained by the addition of formic acid. On completion of the tannage, the hides are piled down in the usual manner for a day or so before further processing.

Chrome Pretannage

(1) Deliming and chrome tan in drum
    Wash: 30–40 minutes
    Deliming with ammonium sulfate, sodium bisulfite
    Pickle: salt and sulfuric acid
    Chrome tan: 1% chrome tan powder
    Run and tan overnight

(2) Wash out of drum and tan with gradually increasing
    concentration of Mimosa. Run until penetrable.
    Add sodium bisulfite pH adjustment to 3.2 to 3.3.
    Penetration obtained in about 22 hours.

LOOKING AHEAD

There are several dominant factors in the trends of the vegetable tanning industry. First is the availability of vegetable tanning extracts. The loss of the chestnut wood extract in the southeastern United States has brought about a major shift in the materials used toward the imported quebracho and wattle extracts. In the future we can expect an eventual decrease in the availability of quebracho extract coming from South America. The vegetable tanning extracts of the future will probably be more from plants which can be produced quickly, such as wattle extract from South Africa and Australia. A second important factor is the cost of labor in harvesting the vegetable tanning materials. As the standards of living have increased, some materials may no longer be economically harvested.

Shortly after World War II the United States government adapted a policy of stockpiling a large amount of industrial materials. This was to make our country less dependent upon imports during any future national emergency. There are considerable quantities of wattle and quebracho extract now in government warehouses. This material is gradually being sold. The availability of these materials will be a factor in the supply of vegetable tanning materials in the United States for the next few years.

A wide variety of synthetic tanning materials have been developed, and their characteristics can be controlled within close limits. The cost of producing these materials should decrease relative to the costs of the vegetable tannins. We can, therefore, expect a gradual decrease in the use of vegetable tanning materials and a corresponding shift in favor of synthetic materials. The need for high-speed production and changes in the standards of leather quality should favor the use of drums. The rather dramatic efforts to establish solvent tannage systems based on dehydration of the skin by acetone have not, to date, shown themselves to be commercially profitable enough to warrant the investment in equipment.

The use of vegetable tanning materials to fill sole leather will probably become less and less significant in the future. If the leather can be impregnated with a polymerizable material then polymerized in situ a new class of leathers will develop. The technical and economic possibilities appear to be favorable.

REFERENCES

BASF, "Tanners Manual—Vegetable Tanning," Ludwigshafen Am Rhein.

Gustavson, K. H., "The Chemistry and Reactivity of Collagen," New York, Academic Press, 1956.

Gustavson, K. H., "The Chemistry of the Tanning Process," New York, Academic Press, 1956.

Herfeld, H., Otto, J., Oppelt, M., Haussermann, E., and Rau, H., *Das Leder,* **16,** 201 (1965).

"Hide and Leather and Shoes Encyclopedia of the Shoe and Leather Industry," Chicago, Hide and Leather Publishing Co., 1941.

J. Am. Leather Chemists' Assoc., Supplement No. 7, Report of the Symposium on Retannage of Chrome Tanned Leather (1960).

J. Am. Leather Chemists' Assoc., Supplement No. 8, Report of the Symposium on Vegetable Tanning (1961).

Kay, A. N., "The Chemistry and Technology of Leather" (O'Flaherty, Roddy, and Lollar, ed.), Vol. 2, p.161, New York, Reinhold, 1958.

"Leather Chemicals," Philadelphia, Rohm & Haas Company, 1955.

Lollar, R. M., "The Chemistry and Technology of Leather" (O'Flaherty, Roddy, and Lollar, ed.), Vol. 2, p. 201, New York, Reinhold, 1958.

McLaughlin, G. D., and Theis, E. R., "The Chemistry of Leather Manufacture," ACS Monograph 101, New York, Reinhold, 1945.

Orthman, A. C., "Tanning Processes," p. 185, Chicago, Hide and Leather Publishing Co., 1945.

Rodgers, A., "Practical Tanning," New York, Baird, 1922.

Sohn, A. W., *Das Leder,* **18,** 102 (1967).

Stiasny, E., "Gerbereichemie," Steinkopff, 1931.

Tanning Extract Producers Federation, "A Survey of Modern Vegetable Tannage," London, 1974.

White, T., "The Chemistry and Technology of Leather" (O'Flaherty, Roddy, and Lollar, ed.), Vol. 2, p. 98, New York, Reinhold, 1958.

Wilson, J. A., "The Chemistry of Leather Manufacture," Vol. 1, New York, Reinhold, 1928.

Wilson, J. A., "The Chemistry of Leather Manufacture," Vol. 2, New York, Reinhold, 1929.

RESINS, SYNTANS, AND ALDEHYDE TANNAGES

10

The syntans, resin tannages and aldehyde tannages have presented to the research chemists, particularly those of the chemical industry, a tremendous opportunity. With the development of the chemical industry and the knowledge of organic synthesis, it became possible to build synthetic molecules to introduce specific characteristics into leather. The synthetic tannins are to the vegetable tannins as the azo dyes are to the dyewoods. They are much better in their specific activity, have a much broader versatility, and their reactions are predictable and controllable. The use of synthetic tanning materials has increased particularly in recent years, and new syntans are developed for the tanning industry each year. The synthetic tannins are useful to the tanner to obtain special effects in processing or leather quality, including:

(1) Clarification of vegetable tan solution

(2) Pre-tanning for faster vegetable tannage

(3) Lightening of color in vegetable leather

(4) Lightening of color in chrome tanned leather

(5) Producing fullness of feel

(6) Producing soft open tanning effect

(7) Mordanting leather for dyes

(8) Aiding in the penetration of dyes

(9) Aiding in shrunken grain effects

The syntans available to the leather industry are constantly in a state of change. Competition for the market of leather chemicals has kept research programs active in both large and small chemical companies. As a need is recognized, new products and formulations are introduced. The exact nature of these proprietary products is seldom revealed for competitive reasons. The syntans and their synthetics, as outlined in this chapter, represent the history and general principles of syntan production and use.

NAPHTHALENE SYNTANS

Naphthalene syntans are the simplest of the syntans and are the easiest to make. Naphthalene is a solid, aromatic compound with a melting point of 80^0 C. The material is melted and to it is added sulfuric acid for the formation of the naphthalene sulfonic acid. Naphthalene sulfonic acid will be in

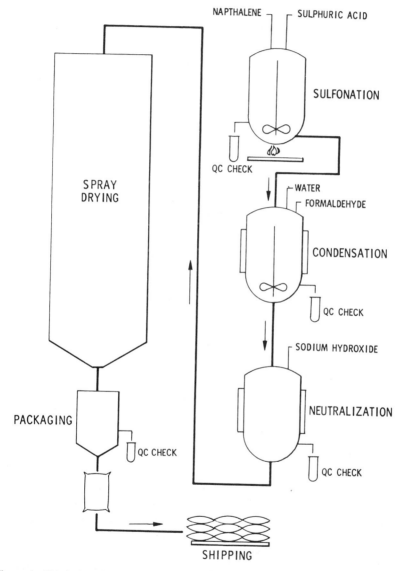

Figure 1 The industrial production of a naphthalene syntan is done in batch process. The naphthalene is charged into a reactor, melted, and sulfonated. The sulfonation mass is then dissolved in water and condensed with formaldehyde. The formaldehyde rate of addition is slow and is used to control the reaction. After condensation, the product is neutralized, if desired, and dried prior to shipping. Frequent chemical analysis is made for quality control (QC check).

173

either the alpha or beta position, depending upon the temperature of the sulfonation. The sulfonic acid is diluted with water and condensed with formaldehyde. The condensation with formaldehyde will bind some of the unreacted naphthalene, and a completely water-soluble product is obtained. The material is strongly acidic. The product may be marketed as an acid liquid for use in whitening chrome leathers or bleaching vegetable tannins. It may also be neutralized to form a neutral salt of the syntan which may be marketed as a liquid; more often, however, it is dried and a solid syntan is obtained. Since the naphthalene sulfonic acid formation is not necessarily 100% completed, the free sulfuric acid in excess from the sulfonation will be present, and most naphthalene syntans will contain some neutral sodium sulfate.

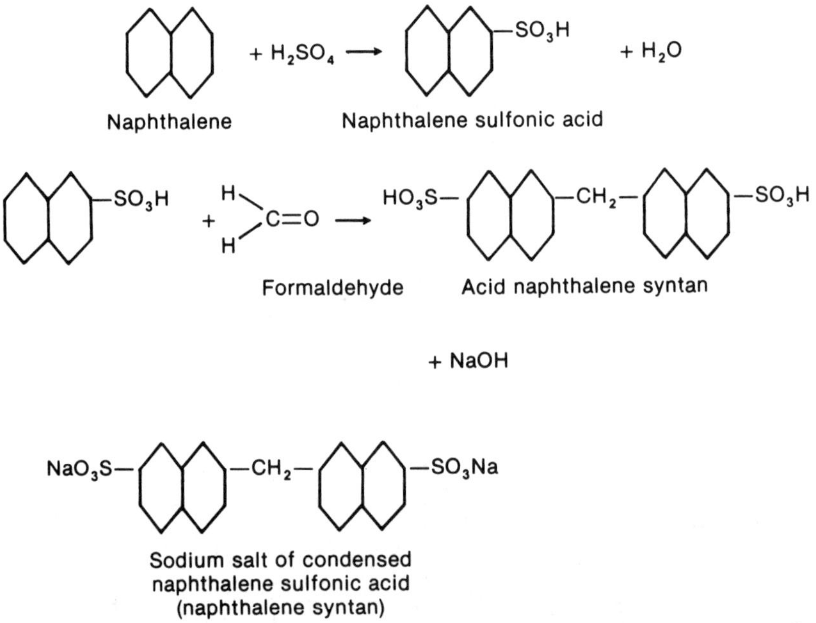

$$\text{Naphthalene} + H_2SO_4 \longrightarrow \text{Naphthalene sulfonic acid}(-SO_3H) + H_2O$$

Naphthalene   Naphthalene sulfonic acid

$$(-SO_3H) + \underset{H}{\overset{H}{>}}C=O \longrightarrow HO_3S--CH_2--SO_3H$$

Formaldehyde   Acid naphthalene syntan

+ NaOH

$$NaO_3S--CH_2--SO_3Na$$

Sodium salt of condensed
naphthalene sulfonic acid
(naphthalene syntan)

Naphthalene syntan·is an example of a material capable of being absorbed by the hide protein through hydrogen bonding. Fixation follows the acid binding curve of the collagen. At a pH near the neutral there is little fixation. As the pH is lowered, the syntan is bound in a classic curve. The naphthalene syntan is a large organic molecule containing hydrophilic sulfonic acid groups. It is, therefore, somewhat related to the detergents. The sulfonic acid group pulls the molecule toward the water, whereas the aromatic naphthalene rings will be attracted to other less soluble materials dispersed in the solution. When dissolved with the vegetable tanning materials, the syntan adds negative charges to form a more negatively

Figure 2   Sulfonators used for the sulfonation of naphthalene in the manufacture of naphthalene-type syntans. *(Courtesy Dewey and Almy Divison, WR. Grace Co.)*

charged colloid. This addition of negative charges to the vegetable tannin will aid in dispersing the tannin particles and help in the penetration of the hide.

The naphthalene syntan will also compete with the vegetable tannin or with acid dyes for the absorption positions on the hide protein, and consequently, it will aid in the brightening of the color and the penetration of dyes. Naphthalene syntan may be looked upon as being a colorless dye when used in a dyeing application.

Since the fixation of the naphthalene syntan is by an acid absorption phenomenon, the reaction is not permanent and the syntan can be displaced. It does not have leather-forming properties in that the leather, when treated with a naphthalene syntan, will not dry soft but will continue to dry hard and bony. There is no real permanent stabilization of the hide fiber and only a slight raising of the shrink temperature. Although the term "tannin analysis" is often applied to the naphthalene syntans in determining their strength, there is no tannin in naphthalene syntans and they have no tanning ability.

EXCHANGE SYNTANS

The exchange syntans are synthetic tannins which are related to the natural tannins in that they contain phenolic hydroxyl groups, and as such, they have the ability to react with the hide protein and produce leather. The first successful development of a commercially feasible synthetic tanning material capable of producing leather was by Stiasny. His first patent, issued in 1911, dealt with the condensation products of phenol or creosol sulfonic acids through the use of formaldehyde. The synthesis is similar to the synthesis procedure outlined for naphthalene syntans above. This simple method points out some of the basic principles in the manufacture of an exchange tannin.

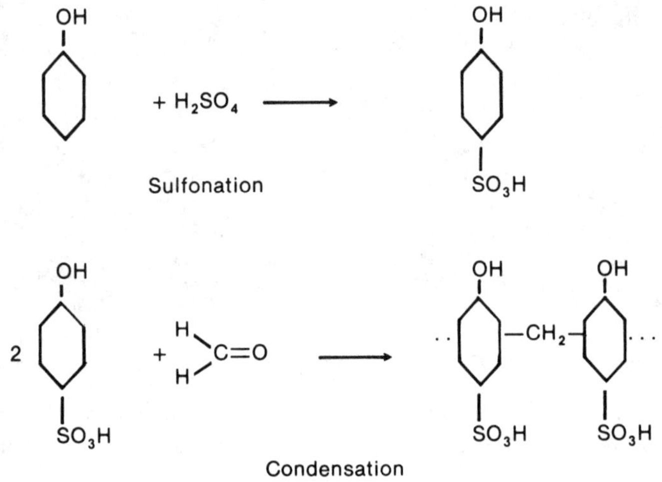

Sulfonation

Condensation

When phenol is condensed with formaldehyde, a novolac resin is formed. This is a thermal setting resin, and its hardness and molecular weight depend upon the ratio of formaldehyde to phenol. At molar ratios greater than one part formaldehyde to one part phenol, the theoretical molecular weight is infinite and a hard resin results. With the molar ratio adjusted properly, so that an average molecular weight is about 300-400, a thick, syrupy material is obtained. This is insoluble in water and, for practical application, leather water dispersion is necessary. The resin is sulfonated by the addition of sulfuric acid. The route to the final product, a sulfonated phenol formaldehyde resin, can be either by sulfonation followed by condensation or by condensation followed by sulfonation.

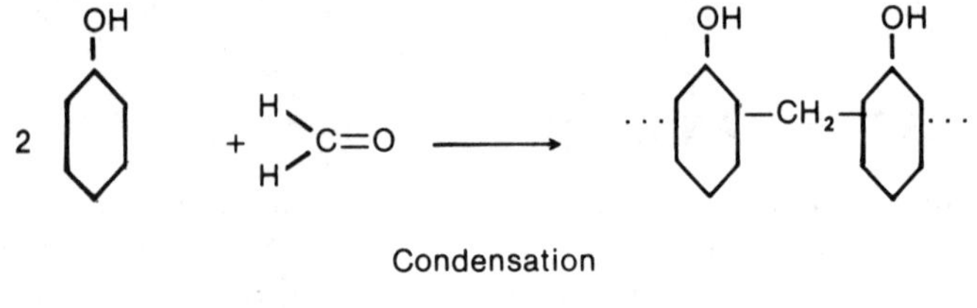

Condensation

$$\cdots \underset{\text{OH}}{\bigcirc} -CH_2 - \underset{\text{OH}}{\bigcirc} \cdots \; + \; H_2SO_4 \longrightarrow \cdots \underset{\text{OH}}{\bigcirc} -CH_2 - \underset{\substack{\text{OH} \\ SO_3H}}{\bigcirc} \cdots$$

Sulfonation

The mole ratio of the condensing agent (formaldehyde) is important. The more formaldehyde, the higher the average molecular weight.

Mole ratio	Average number of phenolic nuclei per molecule	Approximate molecular weight
.5	2	300–350
.66	3	450–500
.75	4	600–700
.80	5	750–900
.9	10	1500–2000
1.0	%	%

If the molecule is too small, there will be poor tanning action. If the molecule is too large, there will be poor penetration into the leather.

The quantity of sulfuric acid, or the degree of sulfonation, also affects the tanning properties. Sufficient solubilization is needed to maintain a true solution, but excess sulfonation decreases tanning efficiency.

The early work of Stiasny was spurred on for commercial application due to the isolation of Germany from sources of vegetable tannins during World War I. With the commercial practicality of exchange tannins well established, the research continued and a wide variety of exchange tannins were developed, particularly in Germany.

Two factors in the exchange tannins are of prime consideration: the cost, and specific characteristics. With phenol as a starting material, the manufacturing costs make it impossible to make a tanning agent competitive with natural vegetable tannins. If, however, there are incorporated into the tanning material lignosulfonates or naphthalene sulfonic acid materials, the cost per "tan unit" goes down and a more competitive product results. The leather-making properties of the syntan are also greatly altered.

One of the most successful of the lignosulfonic acid types was the Tanigan Extra A which is described as an exchange tannin based on dihydroxydiphenylsulfone, sulfite cellulose and the sulfone formaldehyde resin. As a tanning agent it was successful and was capable of producing leather similar to that obtained from the vegetable tannins. The cost was reasonable and large tonnages of the material were used, particularly during World War II in Germany.

Tanigan Extra A

The nature of the linkage between the phenolic rings is of importance. If Bisphenol A is used, a product of greater light stability is obtained than would be obtained with the phenol formaldehyde condensate. Dihydroxydiphenylsulfone is also usable to produce a more light-stable material. Bisphenol A and dihydroxydiphenylsulfone are, of course, much more expensive than phenol, and their acceptance is predicated in color and better-tanning action.

The type of sulfonation is of importance, since the tanning characteristics as a function of pH depend upon the method of solubilization. Ideally, an exchange tannin could contain no sulfonic acid, and consequently, by solubilization through additional phenolic hydroxyl groups greater tanning ability would result. Several exchange tannins with excellent tanning properties have been made based on resorcinol, but the high cost has precluded their widespread use.

A large number of different exchange tannins are presently being offered to the leather industry based on combinations of dihydroxydiphenylsulfone, Bisphenol A, direct sulfonation, co-condensation with naphthalene sulfonic acid, and other systems. As the molecules become more complicated, more specific tanning properties can be built into them. Examples of the types of materials of various syntans presently held under patents by various manufacturers are described below. These examples are chosen to illustrate different types of products. The quotations concerning the performance of the products are excerpts from the patents.

(a) Syntan, United States Patent 2,129,554; Rohm and Haas

In this patent dihydroxydiphenylsulfone is condensed with urea and formaldehyde to form an insoluble resin. The material has been dispersed with naphthalene sulfuric acid. This early patent 1938 achieves light fasteness through the use of urea formaldehyde condensation and also dispersed of the resins by means of a separate material. The practical patent claim is, "When applied to bleaching the chrome tanned leather these materials give a rapid and intensive surface retannage and an improved quality white leather that has a remarkable resistance to sunlight."

(b) Syntan, United States Patent 2,320,678; E.M. Tassel of France (1939)

This patent is an example of solubilization by omega-sulfonation. The addition product of formaldehyde and sodium sulfite react with the aromatic nucleus to form a soluble product. Condensation with formaldehyde forms the syntan. Properties of the syntan will depend on the proportion of formaldehyde bisulfite and phenolic compounds used. The benefits in the system are as outlined below. "The tannins which are obtained according to the present invention possess the following properties:

(1) the degree of polymerization and the proportion of non-tanning substances may be controlled;

(2) the swelling power may be controlled according to the duration of the reaction;

(3) the tannins give leathers in which the fibers are well open and which are white or light colored;

(4) the solubility may be controlled at will according to the duration of the reaction and the final pH."

(c) Syntan, United States Patent 2,809,088; G. Mauthe (1957)

The patent by Mauthe is one of a series involving a cationic exchange syntan which is a distinct departure from normal practice.

(d) Syntan, Example from United States Patent 3,029,212; Nopco Chemical Company (1962)

The formula given is an example of a solubilization made by condensation of dihydroxydiphenylsulfone and the sulfonated acid of Bisphenol A. The condensation is made by means of formaldehyde. "They are useful as re-tanning agents for chrome tanned leather and as pre-tanning agents to form shrunken grain effects. These novel agents have good filling properties and are light fast. Leather so treated is very plump and has a nice grain. Such leather is further characterized by having higher tensile and stitch tear strengths when compared with leather treated with prior art tanning agents."

The field of the exchange tannins is one that is rapidly changing. Chemical manufacturers are becoming more skilled in the application of the art and are now in a position to synthesize products and build in specific properties for specialty leathers. The aim at present is not to replace the vegetable tannin but rather to make a specialty-chemical type compound which will produce desired qualities in the leather.

Since a particular manufacturer may have several patents or may be manufacturing a commercial syntan without the benefit of patent protection, commercial products being offered are probably not as simple as the products described in the patent literature.

RESIN TANNAGES

The resin tanning agents are materials applied to leather in the form of the methylol derivatives of organic nitrogen compounds.

The simplest of these is dimethylol urea. One of the more commonly used is the methylol derivative of melamine, and one of the more successful is the dicyandiamide resins. These compounds have many of the characteristics of an aldehyde tannage in that the methylol derivative group is a reactive site

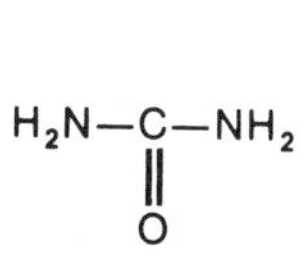

$$H_2N-\underset{\underset{O}{\|}}{C}-NH_2$$

Urea

$$HO-CH_2-\underset{\underset{}{\overset{H}{|}}}{N}-\underset{\underset{O}{\|}}{C}-\underset{\overset{H}{|}}{N}-CH_2OH$$

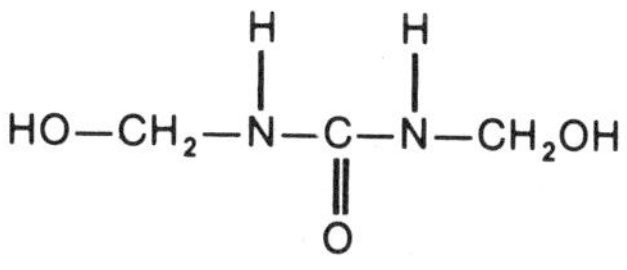

dimethylol urea

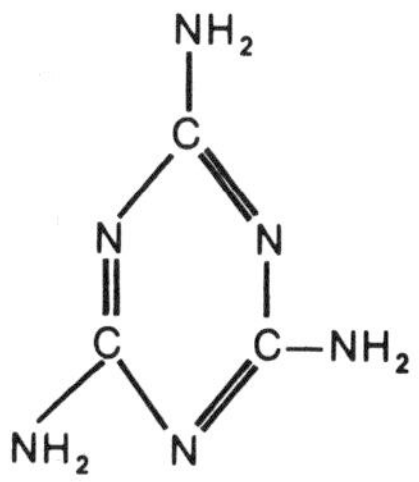

Melamine

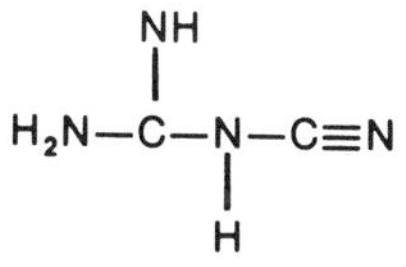

dicyandiamide

and reacts with the amino group on the protein in an aldehyde-type condensation. The organic nitrogen compound component of the product builds up the molecular weight and controls the characteristics of the tanning material with respect to fullness and firmness. The dicyandiamide resins are used in conjunction with the vegetable tanning materials or as a substitute for them in a re-tannage of chrome leather. The compound has a positive charge and therefore can be attracted to the leathers containing a negative charge due to the presence of vegetable tannin or the anionic synthetic tannins of the naphthalene or exchange type. The ability of the resin tannins to combine with the leather builds up large molecular weight compounds in situ and their value is primarily in the area of improving cutting value by the fixation of the resin in the flank areas. In this respect a small quantity of the nitrogen-type or resin-type tanning agents has the same effect as a much larger quantity of vegetable tanning material. The resin tanning agents have the advantage of being colorless and light-stable and have good application in white leathers. Their use is not as widespread as the naphthalene syntans or the exchange tannins, but they do have their place in a wide variety of leathers.

Dialdehyde starch

The United States Department of Agriculture, in its research on chemical uses of agricultural products, developed a method of oxidizing starch to a dialdehyde product, or oxystarch product, which has the ability to react with protein to produce a tanning effect. The availability of the dialdehyde starch at a modest cost indicated the possible use of this material as a substitute for the vegetable tannins. Extensive research has been undertaken and successful leather has been produced with this material. The quality of the leather produced and the cost of the raw materials has precluded its commercial acceptance on a large scale. However, in the future, with changes in labor costs of harvesting vegetable tanning materials, it could very well develop as a major leather-making material.

Formaldehyde

Formaldehyde has been used as a tanning agent and as a means of hardening proteins for many years. It is a remarkable tanning material that usually has its practical application as an adjunct to some other tanning agent.

Formaldehyde is the smallest of the aldehydes and has the ability to react with many organic materials having an active hydrogen atom. The reactions of formaldehyde in the manufacture of the syntans given previously are typical. Untanned hide proteins also have many possible reaction sites due to their active hydrogen atoms. The predominant reaction between formaldehyde and hide protein is generally accepted to be with the ϵ amino group of the basic amino acid, lysine. This is a typical amine-formaldehyde reaction with the formation of the methylol derivatives.

$$R-NH_2 + \begin{array}{c} H \\ \diagdown \\ \diagup \\ H \end{array} C=O \longrightarrow R-NH-CH_2-OH$$

This reaction can further proceed with the reaction of another amino group to form a condensation reaction.

$$R-NH-CH_2-OH + H_2N + R \longrightarrow R-NH-CH_2-NH-R$$

In tanning this would result in a cross linking of the protein and a stabilization of the tannage. Due to stearic considerations, it is generally conceded that only a small portion of the formaldehyde fixed could be involved in cross linking.

The reaction of formaldehyde with an amine to form a methylol compound is favored by the presence of the amine in the uncharged state.

$$R-NH_3 + \rightleftharpoons RNH_2 + H^+$$

For this reason aldehyde tannage reaction takes place faster and to a greater extent at high pH. Practical aldehyde tanning is best conducted in most cases at a pH range at 6.0 to 8.0.

Formaldehyde alone when reacted with hide protein can raise the shrink temperature as much as 32° C. In this respect formaldehyde is the best of the simple aldehydes. When used in combination with other tanning agents formaldehyde may have an additional tanning effect. With the mineral tanning materials it may be used either as a pre-tanning agent or as a re-tanning agent. The latter method is usually preferred. Formaldehyde will increase the shrinkage temperature of most mineral tanned leathers. The effect is of little significance in chrome tanned leathers but is regularly employed with alum tannages for glove leathers and fur tannages.

The re-tannage of vegetable tanned leather by the use of formaldehyde may result in an increase of the shrinkage temperature by as much as 17° C. The formaldehyde can also react with the vegetable tanning materials in the leather. Formaldehyde re-tannage of vegetable tanned insoles will increase their resistance to perspiration.

GLUTERALDEHYDE

Gluteraldehyde is an excellent tanning material which is capable of cross-linking proteins due to its bifunctional nature. When used as a tanning agent in itself, it can produce a good leathering effect and has a dark yellow color. Through the Department of Agriculture research, the application of the gluteraldehyde as a supplementary tannage to chrome tannage has resulted in a greater stabilization of the protein fiber which results in a raising of the shrink temperature and a greater resistance of the leather to chemical attack. The gluteraldehyde tanned shearlings have been found to have excellent resistance to urine and are washable. These properties of chemical resistance and washability of the leather have led to the widespread use of this type of shearling in hospitals.

For the tanning industry, the increased chemical resistance of the gluteraldehyde points to the increased use of this material, particularly in areas where chemical resistance is desired. This includes increased perspiration resistance on insoles, shoe leathers, and linings, and also the application of the gluteraldehyde for increased resistance to barnyard acids. Gluteralhyde has been found to have a softening effect on leather and also has the ability to make the leather more receptive to subsequent chemical treatments for water-repellency and other specific effects. In an increasing number of leathers, gluteraldehyde is being employed as a re-tanning agent and as a supplementary material to other tannages.

Glutaraldehyde can be applied to leather in the pH range of chrome tanning and in retanning. It is possible to add glutaraldehyde to the initial chrome tannage and obtain the benefit of glutaraldehyde tannage simultaneously with chrome tanning. The two tannages are compatible and independent on pickled stock.

Gluteraldehyde has been widely accepted in the leather industry particularly in North America. It imparts softness and remarkable resistance to perspiration that is of particular importance in glove and lining leather. Most chrome leather tanners use gluteraldehyde in at least one general formulation. Some tanners use gluteraldehyde in almost all their leather.

LOOKING AHEAD

If we include as synthetic tannins all materials of synthetic origin capable of combining with hide protein, either in a primary tannage or a re-tannage system, the number of materials available is fantastically high and far greater than can be touched upon, even in a preliminary way, in a book of this type. We are gaining very rapidly in our knowledge and understanding of the method of application and combinations of such materials. We also have available new and different methods of application which open increased possibilities.

The development of syntans has been almost entirely a chemical industry effort. This effort in practical research is in anticipation of industry use of the products. Some major chemical companies have concluded that this effort is not cost-effective and have abandoned the leather industry as a viable customer. The cooperation of the tanner with his suppliers in new product development is essential in this and all other leather chemical developments.

DYEING

11

The vegetable tanning process, common to practically all primitive leathers, imparted a natural brown color, the shade of which depended upon the extracts and the type of tanning process employed. The quantity of oil used in the leather gave it deeper colors so that heavy boot leathers were dark brown and saddle leathers containing less oil were a lighter brown. The introduction of a different color was a very difficult and expensive process and was restricted to luxury items. The color of vegetable tanning materials was of great importance, and the lighter-colored tannins were premium. The leather industry still is affected by the traditions of the past so that natural vegetable tanned colors (the browns and tans) are the most popular shades.

With the beginning of commerce to the western hemisphere, the brightly-colored dyewoods of the trees of the Caribbean area led to the traffic in logwood and its related compounds. With the discovery of mauve by Perkin and the subsequent development of the synthetic dye industry in Europe, a wide variety of colors became available for the leather industry, and a whole new technology was born. Today, with chrome tanning and a wider range of tanning materials, practically any color desired can be made on any type of leather. Wood extracts and aniline dyes are all used to obtain desired color effects in modern leather production.

The dyeing of leather presents some unique problems. In the three-dimensional matrix of fibers characteristic of leather, penetration of the dye is of great practical importance. Leather that is to have very gentle use only, such as upholstery leather, can be dyed on the surface; there is no need to expend valuable dyestuffs in an attempt to obtain deep penetration. In the case of a garment suede or shoe suede leather, however, the constant scuffing would result in light spots unless the dye were deep in the leather fibers. In some cases where the leather is to be be finished, such as in shoe uppers, the dye gives a colored base so that scratches in the finish do not appear as undesirable blemishes. It is evident that the degree of penetration of the dye is of great importance and must be carefully controlled.

Since leather is a protein material, the chemical characteristics of the fiber are greatly altered by tanning. Fixation of the dye will be greatly altered by the tannage and by the presence of other materials attached to the protein.

186

The presence of oils in the leather and the application of other materials in subsequent processes may affect the color shade and the permanency of fixation. Therefore, dye fixation in leather is a balance of a number of phenomena, and the skillful application of dyes by the tanner is a complicated combination of science, technology, and art.

The dyewoods were closely related to tannins and behaved in a manner somewhat similar to the vegetable tanning materials, both in fixation to the leather and in methods of manufacture. The principal dyewoods are logwood, fustic, osage orange, and hypernic. At the present time only logwood is utilized.

Logwood is the heart wood of the tree known as Hematoxylin Campechianum, which grows abundantly in the West Indian islands of Haiti and Jamaica. The wood is stripped of its bark and shipped in conveniently sized logs to the extract plant where it is extracted in much the same manner as vegetable tannins. The extract is concentrated by evaporation, and finally a crystalline compound is obtained. The color of the logwood will depend upon pH and will be intensified by the addition of an alkali. Logwood has been extensively applied in the development of base colors in upper leathers, particularly calf. The crystalline material is dissolved, added to the leather in a drum, and the color is developed in the leather by the addition of a "striker." The color developed will depend upon the dyewood used and upon the striker. The proper selection of dyewood and striker will yield a variety of colors.

Development of Color with Logwood

Striker	Color
lead acetate	dark brown-violet
ferric sulfate	black
copper sulfate	dark red-violet
tin	red-violet
sodium bichromate	black
potassium titanium oxalate	brown

It was common practice up until a few decades ago to apply the dyewoods and their strikers to give the base colors to the leathers. The final shading was adjusted by the addition of aniline dyes as part of the finishing operation. The dyewoods, however, do not allow the broad range of colors desired in modern fashions.

THE SYNTHETIC DYES

The synthetic or coal-tar dyes are classified according to their composition into various groups of which azo dyes are the most important.

Acid dye: Acid Blue 2B

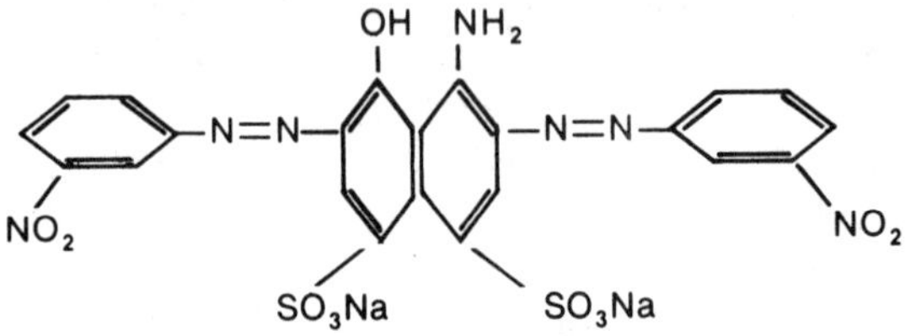

Note the conjugated double bonds C=C—C=C—N=N—C=C— in this typical acid dye. The molecule is solubilized by the acid SO_3 groups, making it anionic.

Basic dye: Bismarck Brown G

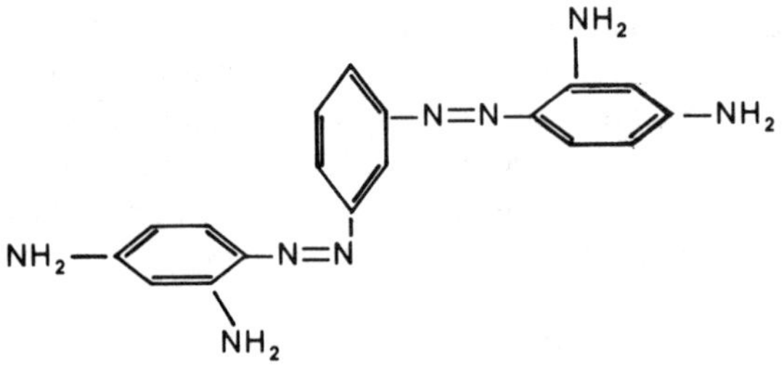

The molecule of this typical basic dye has conjugated double bonds similar to those found in other dyes. Solubilization is by the basic —NH_2 groups, making it cationic.

The intermediate or specific compound being coupled can be varied greatly, and a specific wide variety of colors can be obtained. There are over 1500 commercial dyestuffs, most of which were discovered in the early twentieth century. The azo dyes and the synthetic dyestuffs are characterized by large structures containing a number of conjugated double bonds, i.e., in each dye molecule the carbon atoms have alternate single and double bonds. A conjugated double bond structure is associated with the transference of electrical charges from one end of the molecule to the other. This is a factor in the fixation of the dye and in the development of color.

In the manufacture of azo dyes, coupling takes place in a water solution, and after coupling the solution is acidified or neutralized; then, salt is generally added and ice introduced into the vat containing the coupled dye. Under the conditions of high salt content and low temperature the dye will precipitate from the solution and can then be removed by means of a filter press. The dye is then dried and standardized by measuring its coloring value. From the measured colored value, the dye is blended with salt to obtain the standard color of that particular grade and type.

The azo dyes may be classified as acid, basic, or direct dyes. The direct dyes are so called because of their ability to dye cellulose or cotton material directly without the addition of a mordant. The acid dyes are anionic materials due to the presence of sulfonic acid groups on the dye and their predominantly acid character. The basic dyes are cationic in nature, with free amino groups on the dye molecule.

THEORY OF DYE FIXATION

The dyewoods, being closely related to the vegetable tannins, are similar to the vegetable tannins in their theory of fixation. The dyewoods are less colloidal in nature and will penetrate more easily than the vegetable tannins, but the same general principles are involved. Fixation is by hydrogen bonding, and the lower the pH, the greater the fixation.

Direct dyes are attracted to leather fiber and will be absorbed on the surface of the leather by physical forces rather than by strong chemical and physical bonds. Direct dyes are not penetrating when used on chrome tanned leather; rather they will be absorbed on the surface.

Acid dyes behave as weak acids, being absorbed by the hide. The fixation and penetration of an acid dye follows the same general laws that apply to the acid syntans. The acid dyes are attracted to the leather through the positively charged groups of the hide; therefore, the acid group will be attracted to the amino group on the hide and will be fixed by hydrogen bonding. At low pH the acid dyes will be fixed more readily than at high pH, but they will penetrate deeper into the skin as the pH value increases. Since dyestuffs vary in molecular size, in their degree of solubility, and in their acid-base characteristics, fixation of the dye by the leather will depend on all these factors. Under given conditions the dyes will vary in these characteristics from one dye to another.

The basic dyes are attracted by negatively charged groups on the leather. Under acid conditions leather, like untanned protein, will absorb hydrogen ions and assume a positive charge. The basic dye, also positively charged, will have little affinity for the leather. As the pH is raised, the leather becomes more negatively charged and fixation of the basic dyes is aided.

In addition to the pH factor in the fixation of dyes, the presence of other materials on the leather fiber is important to the behavior of the dye in relation to the leather. Chrome tanned leather carries an additional positive charge due to the presence of the cation reaction with the hide. The acid dyes, therefore, will be more strongly attracted to chrome leather and will fix directly to it. Basic dyes, on the other hand, have little affinity for chrome leather. In order to fix a basic dye onto chrome leather, a negative

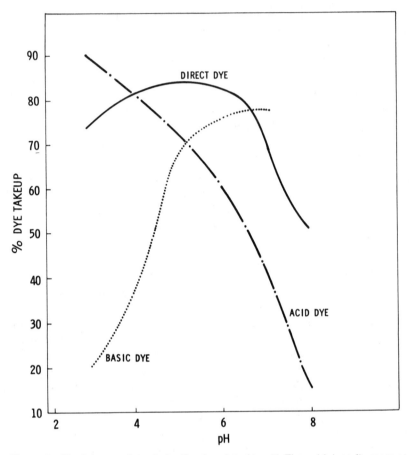

Figure 1   The take-up of dye in leather is related to pH. The acid dyes fix more at low pH. The reverse is true of the basic dyes. A direct dye has high fixation over a wide pH range.

charge must be introduced on the leather. This is done by mordanting chrome tanned leather by adding a vegetable tanning material or an anionic syntan. The basic dye then will be very strongly attracted to the "mordanted" fiber, and stable fixation will take place.

PRACTICE OF DYEING

The application of leather dyes is a batch process that is conducted in a drum. The dye drums are considerably smaller than those used for chrome tanning. The liquor ratios employed are about three pounds of water per pound of blue shaved weight. Even higher liquor ratios may be necessary to maintain even coloring. The comments given on dyeing practices are based

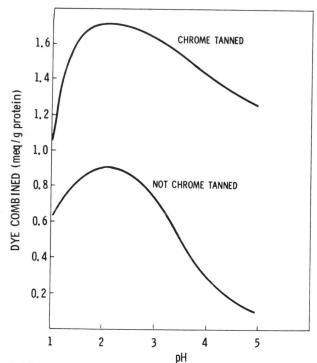

Figure 2 Acid dye attraction to chrome leather. An acid dye will be more attracted to leather that has been chrome tanned. Chrome tanning makes the leather more cationic and therefore more attractive to the minus-charged acid dye.

on drum applications. The same principles apply to both brush dyeng and tray dyeing, as indicated.

Brush dyeing: Brush dyeing is a method of applying dyes to a leather where it is desired to color only one side. It has the advantage of being very economical in dye and is adaptable to obtaining special effects or evenness of dyeing, depending on the skill of the operator. The dye solution is applied to the surface of the leather and brushed out. The operator then watches the application of the dye, and additional dye is applied where needed. Two or three coats may be necessary to obtain the desired shade. To use this method, it is necessary to have dyes of good solubility in order to avoid color streaks from the brushing and from suspended particles of dye. In order to obtain more evenness of color, weak dye solutions are used.

Tray dyeing: Tray dyeing may be employed for the dyeing of small skins. With this system a shallow pan, large enough to take the spread-out skin, is used. Two skins are usually placed flesh to flesh and are lowered into the solution. When the desired depth of penetration of the dye on the grain side

Figure 3   Brush dyeing of light glove leathers. *(Courtesy Seta Leathers Ltd.)*

is obtained, the skins are removed, placed grain to grain, and dipped again to get the desired depth of penetration on the flesh side. Because of the high labor cost, the method is satisfactory for limited production only. Like brush dyeing, dyes are needed which have good solubility at low temperatures and which will attain even fixation without close pH control.

Drum dyeing of chrome tanned leather

The actual practical application of dyes is a complicated process which involves the competition of various reactive materials with each other in their ability to fix onto the hide fiber. In dyeing, chrome tanned leather should be washed thoroughly free of salts to prevent salting out of dyes into the leather. The pH is raised and the dye (acid dye) is introduced. If the pH is above the pK value for the acid of the dyestuff, and the leather is not strongly cationic, the dye will penetrate deeply into the skin and an even fixation of the color will result.

A naphthalene or anionic syntan added with the dye will cause competition between the syntan and the acid dye for the available positions on chrome tanned leather. This will cause leveling of the shade and greater dye penetration. Vegetable re-tannage of chrome tanned leather will also aid dye penetration. The vegetable tanning materials are not so strongly anionic as to prevent the fixation of acid dyestuffs, but they do have a leveling effect. Acid dyestuffs may be applied with vegetable tanning materials in the re-tanning operation to give the desired base colors to the leather. Once an acid dyestuff has penetrated, pH may be lowered by addition of a weak acid, usually formic, and the reaction between the leather and the dyestuff is completed. Even at this point fixation of the dye may not be permanent. In the subsequent fatliquoring operation the strong sulfonic acid of the oil may result in displacement of the dye from the leather. True color fastness of leather is not attained to the same degree as in the simpler textile system due to the complicating factors of fatliquors and tanning agents.

Basic dyestuffs will result in deep penetration of color if vegetable tanning materials or anionic syntans are not present; basic dyes may be used at the end of a dye formulation in which acid dyes or dyewoods are used. The basic dye will coprecipitate with the acid dye on the surface, giving a bright, strong shade. This method is often employed in the coloring of bright shades on full grain leather.

The percentage of dyestuff used varies greatly, depending upon the type of leather being made and whether or not the dye to be used is a penetrating or surface type. In the manufacture of side leather, usually only a light to medium color (from less than 1% to 3% dye) is necessary on the surface to give a base color over which the finish will be applied. With suede leathers, deep penetration of the color is necessary, and in order to obtain the desired shades, as much as 15% dye may be necessary.

In choosing a dye for leather, and in making up a dye formulation, it is not enough merely to put together two or more dyes which will give a blended shade in accordance with a color card. Consideration must be given to the relationship between the dye and the leather, and dyes which are similar in

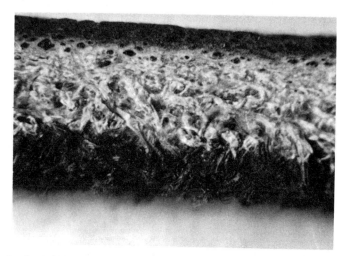

Figure 4. Cattlehide leather in cross section showing differences in fiber structure from grain to the center of the skin. Dye penetration can be observed from both grain and flesh sides.

their behavior toward the leather must be use. It would not be proper, for example, to attempt to make a green leather using a blend of yellow and a blue dye. Under this system, if there were a significant difference in the degree of penetration of the dyes, the dye solution might have the desired green color, yet the leather might be too blue on the surface and too yellow immediately below the surface. In order to obtain the proper color, dye blends should consist of dyes which are close to the desired shade and have similar penetration characteristics. If deep color penetration is desired, all the dyes used should be of the penetrating type for that particular leather. If the dyes involved in the formulation are closely related in their color, variation in degree of penetration from one dye to another will be of less significance in giving variations in shade.

The concentration of the dyestuffs in solution is also of importance. A large volume of water is necessary to maintain proper solutions of the dyes, and in the interest of obtaining level colors, the volume of liquor used in the drum during dyeing is considerably higher than that used in the re-tannage or in the initial chrome tanning.

When the dye is added to the mill it should be completely dissolved. Undissolved dye will be absorbed on the surface of the leather, and streaks may result. Some dyes may undergo decomposition when placed in boiling water; a temperature of about 180° F is usually the highest temperature that is practical for the dissolving of dyestuffs.

The shade of a color on leather will depend upon the concentration as well as the blend of dyes. Leather itself is not a pure white material, so when a dye is added, it must overcome (or "hide") the base color of the leather. In chrome tanned leather, syntans of the anionic type will add an increasingly negative charge to the leather. The naphthalene syntans also cause a formation of a light-colored chrome complex on the leather which contributes to whitening. The acid dyes, competing with the syntans for position on the skin, must penetrate further into the skin before they are fixed, since more of the available sites on the leather are occupied by the acid syntan. The addition of the syntans, therefore, will result in increased penetration and the formation of lighter and pastel shades. The addition of vegetable tanning extracts also results in an increasing electronegativity of the fiber and greater penetration of the acid dyestuffs. The addition of either an anionic syntan or a vegetable tanning agent can act as a mordant for the fixation of a basic dyestuff.

Designation of color

Several systems have been developed in the designation of color. Some of these are applicable to the instrumental methods, but for practical leather application and as an aid in color matching, these systems may be simplified through the form of a color solid. In systems of this type the color wheel takes the form of a circle with the colors red, orange, yellow, green, blue, and violet as sectors within it. In the center is a dark mixture, or black. The identity of the color as red, yellow, green, or blue is called its *hue*. In addition to hue, there is the mixture between the particular color and white or black, i.e., the *shade* may be a light red shade or a deep red. This is represented on the color circle as vertical lines perpendicular to the color circle, with white being in one direction and black in the other. We may designate a point on this color solid to correspond to the shade and hue of a particular dyestuff or a particular leather. Ideally, the color resulting from the blend of this dyestuff with another dyestuff in another position in the color solid can be represented by a straight line between these two points. Practically,

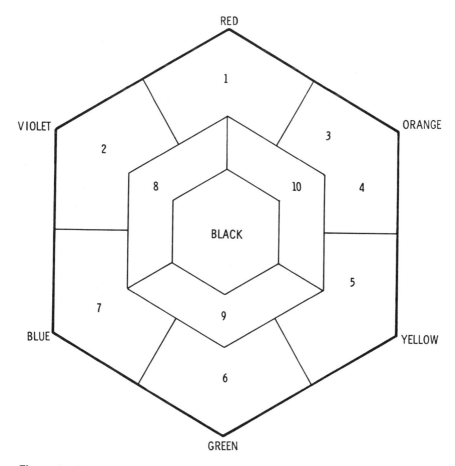

Figure 5 Color wheel corresponding to the acid dyes listed for chrome leather. Similar information is available from dye manufacturers covering a wide variety of dyes.

however, this is seldom accurate, and it may be necessary to add a considerable quantity of one dye to another to shift the color significantly. Such color guides can be used only as a general guide to the choice of other colors to bring about the necessary changes in shade.

In the development of a color or a particular shade desired, successive additions of dye are often made in the experimental dyeing. This may result in a complicated dye formulation with many small additions of dye. Once the total quantities of dye necessary to bring about the desired color have been determined, the formula can often be greatly simplified and the number of dye additions decreased. Care must be taken, of course, to add the dyes in proper sequence and to prevent the simultaneous addition of dyes which may be incompatible with one another.

The various dye structures are described in the color index and relate to specific chemical formulae. These same dyes may be manufactured by several different manufacturers and be described by trade names in the company's literature. These dyes differ in their strength and purity and cannot be substituted on a pound-for-pound basis simply on the basis of their color index number. The characteristics of a few common dyes are given in Table 1 and indicate the type of information available to the tanner. For more complete tables of the individual dyes, the reader is referred to the literature of the dye manufacturers.

TABLE 1

The Properties of Acid Dyes on Chrome Tanned Grain Leather

Listed below are some acid dyes and their properties. These dyes correspond to the numbers given on the color chart. The properties of the dyes are typical of the information that would be supplied by the manufacturer as a guide to the dyer. This list is given only as an illustration and is not meant to be a guide for dyeing. For the solution of practical dye problems a much greater list of dyes and their properties would be needed. The reader should try working out a few theoretical dye problems based on these dyes, the color chart, and the concepts outlined in the chapter.

(1) *Acid Red 73*

Covers well—yellow-red. Can be used on all types of leather.

Levelness	Good
Exhaustion	Fair
Penetration	Good
Glazing	Slightly heavier
Grain and flesh	Similar
Bleeding	Fair resistance to fatliquor

(2) *Acid Violet 54*

Surface dyeing—clear red-violet shade. Most suitable for garment suede in calf or sheep.

TABLE 1 (cont.)

Levelness	Good
Exhaustion	Good
Penetration	Slight
Glazing	Slightly heavier
Grain and flesh	Heavier on flesh
Bleeding	Fair resistance to fatliquor

(3) *Acid Orange 8*

Good penetration—red-orange. Can be used for finishing and drum dyeing on all types of tannage.

Levelness	Good
Exhaustion	Good
Penetration	Good
Glazing	True color
Grain and flesh	Similar
Bleeding	Fair resistance to fatliquor

(4) *Acid Orange 10*

Highly penetrating, good solubility, low tinctorial value. Best on chrome tanned leather. Can be used in all types of dyeing applications.

Levelness	Good
Exhaustion	Fair
Penetration	Good
Glazing	True color
Grain and flesh	Flesh heavier
Bleeding	Very good resistance to bleeding in fatliquor

(5) *Acid Yellow 42*

Slight penetration—red-yellow shade. Grain and flesh even color.

Levelness	Good
Exhaustion	Good
Penetration	Slight
Glazing	Glazes more green
Grain and flesh	Even
Bleeding	Good resistance to fatliquor

(6) *Acid Green 3*

Yellow-green. Suitable for all types of leather and all application methods.

Levelness	Good
Exhaustion	Good
Penetration	Slight
Glazing	True color
Grain and flesh	Flesh heavier
Bleeding	Fair resistance to fatliquor

TABLE 1 (cont.)

(7) *Acid Blue 92*

Best used in drum dyeing to tone the shade of the color.

Levelness	Good
Exhaustion	Good
Penetration	Slight
Glazing	Slightly heavier
Grain and flesh	Slightly greener flesh
Bleeding	Fair resistance to fatliquor

(8) *Acid Violet 3*

Dull violet shade. For drum dyeing on all types of tannages.

Levelness	Good
Exhaustion	Fair
Penetration	Very penetrating
Glazing	True color
Grain and flesh	Flesh darker
Bleeding	Fair resistance to fatliquor

(9) *Acid Green 20*

Yellow-green shade. Rapid fixation on chrome leather. For drum dyeing garment sheep or splits.

Levelness	Good
Exhaustion	Good
Penetration	Good
Glazing	Heavier
Grain and flesh	Similar
Bleeding	Resistance to fatliquor

(10) *Acid Brown 14*

For chrome or vegetable tanned leather.

Levelness	Fair
Exhaustion	Fair
Penetration	Slight
Glazing	Glazes slightly heavy
Grain and flesh	Even color shades
Bleeding	Fair resistance to fatliquor

Measurement of color

The color of an object arises from its characteristic of absorbing light of various wavelengths to unequal degrees. Normal daylight is made up of a wide variety of wavelengths, and in the visible range these wavelengths fall between 400 and 700 nanometers. The human eye is capable of distinguishing

Figure 6 Coloring and fatliquoring wheels. Coloring and fatliquoring is done as part of a continuous multistep operation. For coloring and fatliquoring, the evenness of the color and the even distribution of the oils and retanning materials require a larger amount of water in relation to the hides. Better quality is, therefore, obtained by using small equipment loadings. Coloring and fatliquoring drums are usually about 8 ft. in diameter by 6 ft. wide as shown in this photo, and can accommodate 400 to 1000 lbs. of blue shaved leather at a time. When the coloring and fatliquoring operation is complete, the sides are usually removed, one at a time, by hand and placed on the horses to drain. It is important that the sides be laid out smooth, to avoid wrinkles and lay marks. *(Courtesy Granite State Tanning Co.)*

many different intensities and combination of intensities of these lights; this results in the ability to distinguish between as many as a million separate colors. Several systems of industrial importance have been developed to identify and define these colors. The simplest of these systems was devised by the Commission International for Colorage in 1931. In this system the three quantities of colors, red-green-blue, are described as being present in various proportions to give the sensation of the color indicated. Instruments have been developed which will measure color, based on a tri-stimulus system, and three reflection measurements are taken. This system gives a crude indication of the color but is not adequate to describe the colors of leather sufficiently to be superior to the human eye.

A very sophisticated system of measuring color, which has recently come into practical application in industry, is based on the work of Hardy of the Massachusetts Institute of Technology. A spectrophotometer is used, in which the amount of color reflected at a particular wavelength is measured

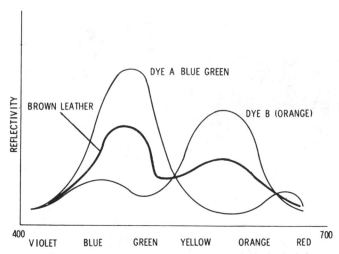

Figure 7   Represented are the reflectivity curves of two hypothetical dyes and a mixture of the two. Color measurements of this type are being used to aid in solving practical dye problems.

over the entire visible range. A reflectance curve for the substance is then obtained. Reflectance curves are also obtained of the same material when dyed or colored by a particular system. Through the use of a computer, a calculation can be made which would indicate the relative amounts of each dye necessary to give a sum total of reflectivity in a pattern similar to the sample being matched. Sixteen reference points are used in present equipment, and this system is far superior to the tri-stimulus or matching systems. The computer then calculates a solution to sixteen simultaneous equations and indicates the dye blend desired. In working with textiles, this has been found to be a helpful tool which greatly speeds the ease of color matching. The application of this instrument to leather has been limited so far because of the complexity of the colored surface of leather. The tannage, dyes, and fatliquors, as well as the nature of the leather surface, all play an important part in the color characteristics of the leather; thus far sufficient data has not been obtained on the many different leathers to permit useful applications of this technique. The principles involved, however, should be understood by the leather technologist for the practical understanding of his leather dyeing problems. Individual leather dyes have been investigated thoroughly by the dye manufacturers as to their characteristics on leather of different types. As a result of this long experience, the more common leather dyes are described in the dye manufacturers' literature as to their traits of penetration, compatibility with various tanning agents, and stability to fading and washing out under various conditions.

LOOKING AHEAD

The development in the dyeing of leather followed the intensive work done in the dyeing of textiles. With the development of the synthetic dye industry in the United States, intensive research was conducted for new dyes and new colors. The field has been well worked over and the rate of invention of new dyes has decreased. Since leather is a minor dye customer relative to textiles, it is unlikely that intensive research will be conducted on the preparation of dyes designed especially for leather.

The future changes in the dyeing of leather, therefore, will be in the methods of application. Assuming present dyeing equipment is employed, these developments will be in the area of special dye assist materials to obtain a better leveling of the color. Dye fixatives which will aid in producing stability of color will also be of importance. In the garment leather field color stability to dry cleaning has not been completely solved.

Another possibility is in the use of either a flow coater or a seasoning machine for dye application. The leather in the crust is a neutral color and can be dyed as desired for specific orders. This system has been used extensively for surface tinting but has not been used to any great extent where deep penetration of the dye is desired. This type of application would probably be based on an organic solvent system.

REFERENCES

Casnocha, J.A., "Sandoz Leather Dyes," East Hanover, N.J.
Frank, G.H., "Manufacture of Intermediate and Dyes," New York, Chemical Publishing Co., 1952
Garverick, K.E. and Leach, R.H., "Leather Dyeing," American Cyananides Co., Technical Bulletin, 1954.
Garverick, K., Jr., "The Dyeing of Leather," "The Chemistry and Technology of Leather," (G. O'Flaherty, Roddy, and Lollar, ed.), Vol. 3, pp. 1-15. G. Otto, Vol. 3, pp. 16-60.
Gustavson, K.H., "The Chemistry of the Tanning Process," New York, Academic Press, 1956.
"Hide and Leather and Shoes Encyclopedia of the Shoe and Leather Industry," Chicago, Hide and Leather Publishing Co., 1941.
Kirk, S.J., "The Chemistry and Technology of Leather," (O'Flaherty, Roddy, and Lollar, ed.), Vol. 3, p. 1, New York, Reinhold, 1962.
Leach, R.H., "The Dyeing of Leather," Roddy and Lollar.
"Manual for the Leather Trade," New York, General Dyestuff Corp., 1950.
Otto, G. "The Chemistry and Technology of Leather," Vol. 3, p. 15, New York, Reinhold, 1962.
Stevens, J.E., "A Tale of Two Trees," American Dyewood Co., 1948.
Stiasny, E., "Gerbereichemie," Steinkopff, 1931.
Wilson, J.A., "The Chemistry of Leather Manufacture," Vol. 1, New York, Reinhold, 1928.

FATLIQUORING

12

The process of tanning leather, whether by chrome tannage, vegetable tannage, synthetic tannins, or the aldehydes, has the primary purpose of preserving the fiber structure from bacterial attack. The preliminary processes of unhairing and bating remove most of the natural oils from the skin. Whatever the course of the pretreatment and the tannage, the leather at the time of the completion of the tannage does not contain sufficient lubricants to prevent it from drying into a hard mass.

Prior to the invention of chrome tanning, vegetable tanned leathers were lubricated by rubbing oils and natural fats into the leather; these coated the fibers, brought about lubrication, and introduced a certain degree of water repellency. This operation was called currying and was a major segment of the tanning industry of the day. The use of oils and natural fats at high levels resulted in dark colors; light-colored leathers were available only in light-weight skins.

The proper lubrication of leather fibers is one of the most important factors in determining the characteristics of leather, both from a utilitarian and an aesthetic point of view. Proper lubrication, or fatliquoring, greatly affects the physical properties of break, stretch, stitch tear, tensile strength, and comfort of leather.

Leather varies greatly in its physical structure not only from hide to hide, but within a single hide. In order to obtain as much uniform cutting value as possible, great care is needed to ensure proper lubrication of the fibers. Over lubrication will result in excessive softness and raggy leather in the bellies and flanks. Under lubrication, or improper penetration, results in hard bony leather that may crack in use.

A fatliquor system that is good for one type of leather may be completely inadequate for another. Adjustments must be made to compensate for the tannage, retannage dyes, and anticipated finishing systems. The skill of the tanner and the distinctive quality of his leather is dependent upon his choice of fatliquoring systems.

LOCATION OF THE OIL

If we consider a cross section of the hide upon bending, we see that on the outside of the bend the fibers must stretch, and on the inside of the bend

202

must compress. In the center of the skin there is very little motion of the fibers over one another during bending. Therefore, in order to avoid excess stiffness of the leather, both the grain and the flesh surfaces must be lubricated, but less lubrication is necessary in the center. The fiber of the skin is much coarser on the flesh side and a very dense pattern of the skin is present on the grain layer. Because of the density of the fibers on the grain layer, the oils may be deposited on the outer surface, resulting in lubrication of the leather at that point and very little lubrication underneath. If this is the case, there may be a tendency of the leather to crack on flexing and stretching, either in the tannery or during the shoemaking operations. If the leather is not properly lubricated near the grain surface, the fibers have a tendency to stick together, and upon flexing, the fiber coalescence can be broken in some areas but maintained in others. This results in a break or grain wrinkle, and subsequent flexing takes place in the same area. The final result is that flexing of the grain occurs with a series of wrinkles relatively far apart rather than very close together, as in properly lubricated fibers. A very poor, or coarse, break of the leather results.

The break of leather is very important in cutting of shoe components. A tight break is one of the most important factors in lasting good appearance in shoes.

A major chemical company (one of the author's consulting clients) conducted an interesting test of leather quality. Using a panel of employees, two pairs of shoes were purchased for each person: one top quality pair and one cheap pair. At the end of two months, after equal use, the cheap shoes had developed very poor break and were out of shape, while the top quality shoes were nearly as good as new. The panelists independently expressed the opinion that the quality shoes were the better investment.

A strongly penetrating oil could penetrate completely in the loose belly and flank areas of the skins, resulting in a very soft, raggy, open leather; yet, the leather may be cracky in the bend areas.

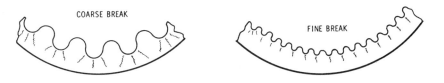

Figure 1 The break of leather. Proper tannage and fiber lubrication will allow the fibers of the leather to move easily and uniformly in relation to each other. Also, the leather will have a fine break which adds to its beauty and durability.

Fatliquors are seldom a single component. Fatliquors are usually emulsion systems consisting of a blend of oils to give a desired degree of lubricity and a balance of emulsifying agents to deposit the oil in the leather as desired.

Proper fatliquoring, or lubrication of leather, is the result of a skillful balance of the oils being used and methods of applying them. In the fatliquoring process, which is the most common method of lubrication, the leathers are treated with emulsions of lubricating oil in a water system. Under the conditions of the fatliquor the emulsion is deposited in the leather with the absorption of the oil on the fibers of the skin. The water is then removed in subsequent drying operations and the oil remains deposited in the leather, lubricating the fibers. To understand fatliquoring, therefore, we must understand the mechanism of the emulsion systems and their behavior.

THEORY OF EMULSION

Let us consider the forces acting upon the molecule of a liquid in a small droplet. If a molecule is located within the droplet, it will be attracted by other molecules of the same material by forces of attraction that are equal on all sides. At the surface of the liquid, however, the forces of attraction will be toward the center of the droplet and toward the molecules adjacent to it at the surface. The forces of attraction for the air, or media in the other phase, will be very small. The net result is that the surface of the droplet will have a tendency to pull inward and decrease its size as much as possible. This phenomenon is known as surface tension. When the droplet is a surface between two immiscible (mutually insoluble) liquids, this tension is referred to as interfacial tension.

If oil is placed in water and is violently agitated, a large number of small droplets will form. As the interfacial tension between the water and the oil is quite large, the oil droplets will gather together to form larger droplets, and

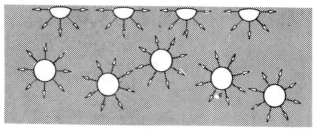

Figure 2   The forces of attraction between molecules in a liquid are even in all directions. At the surface of the liquid the forces are uneven. This uneven balance of forces causes tension and governs the formation of droplets.

eventually a separate layer of oil will result. If some material is introduced into the system so that the interfacial tension between the oil and water is decreased, the rate at which the separating out or "breaking" of the emulsion takes place can be decreased greatly. The forces of interfacial tension can be decreased to such an extent that the emulsion is, for all practical purposes, permanent.

Soaps

The introduction of soap (sodium salt of a fatty acid) into an oil-water system will result in the soap molecule assuming an oriented position at the oil-water interface. The hydrocarbon end of the fatty acid will be attracted to the oil phase, and the carboxyl group of the fatty acid of the soap will be attracted to the water phase. As a result, the interface between the oil and water droplet now is two interfaces, one between the oil and the soap, and another between the soap and the water. The soap is ionized, carboxyl groups are negatively charged, there is very little attractive force between the fatty acids, and the interfacial tension is greatly decreased. The individual droplets also will assume a negative charge, due to the negative charge of the fatty acid of the soap, and will repel one another by electrostatic forces. Under these circumstances the soap is an effective emulsifying agent for oil; a "permanent" emulsion results.

If the pH of the solution were decreased so that the ionization of the carboxyl groups of the fatty acids were repressed, the surface of the water droplets would no longer repel one another and the emusion would break. Soap solutions and soap-water solutions, therefore, are not stable in the presence of even relatively weak acids but are stable in neutral and alkaline pH ranges.

The divalent cations (Ca^{++}, Ba^{++}, Mg^{++}, etc.) also have a specific reaction with soap which is of practical significance in leather manufacture.

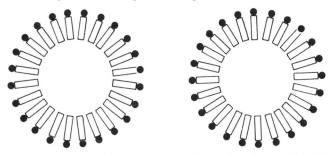

Figure 3   Two oil droplets stabilized by soap. The ionized fatty acids of the soap are oriented with their charged portions toward the water. These droplets are negatively charged and the emulsion is anionic.

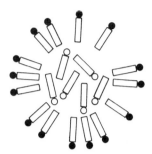

Figure 4 Oil droplet stabilized by soap at low pH. At low pH, the fatty acid of the soap is less ionized. The fatty acids then enter the oil phase and the droplet loses some of its charge and stability.

Calcium ion will be attracted to the soap fatty acid, and a calcium soap will be formed. As calcium soaps have very limited solubility in water (each containing 2 moles of fatty acid per mole of calcium ion), they will be absorbed into the oil molecule and the effectiveness of the fatty acid in dispersing the emulsion will be lost. The calcium ion, therefore, will spew out oil suspensions of soap-stabilized emulsions. The well-known behavior of many toilet soaps, when used in hard water, is an example of this.

In addition to acidity and additional calcium ions, salt may break an oil-in-water emulsion. The electrical charges between the droplets may be counterbalanced by a large number of ions in the water which could bridge the electrical charges between the droplets and result in the precipitation of an insoluble phase. This is known as "salting out" of the emulsion. The same phenomenon is used in degreasing.

Sulfated oils

Leather, particularly chrome tanned leather, is acid and has a pH of about 4 at the time of fatliquoring. At this pH, soap emulsions are too near the precipitation point for proper control. Although soaps are used in some fatliquoring, greater stability at low pH is desired for most applications. This is done by sulfation or sulfonation of the oils. The addition of the SO_3H or SO_4H group into the oil modifies the emulsion characteristics in that the dispersive power of the sulfonated group is much greater than the carboxyl group of the soap. Furthermore, as the pH is decreased, since these are stronger acids, the emulsion is stable in pH ranges as low as pH 3, depending upon the nature of the oil and the extent of sulfation. In modern tanning practice the sulfonated or sulfated oils are by far the largest category of fatliquoring materials presently being used.

The manufacture of a sulfated oil. The techniques of manufacturing fatliquoring oils are as varied as the materials used. As an example of a fatliquor preparation, the following general procedure is given:

The oil is received in tank car quantities, then chilled and filtered. This pressing of the oil removes some of the high melting point components and gives the oil a lower cold test. The higher melting materials are useful in the making of other products. The pressed oil is checked for quality and then sulfated. The sulfation, shown in Figure 5, is by the addition of sulfuric acid. The reaction is exothermic and proper temperature is maintained. When sulfated to the desired level the oil is put in a wash tank. In the wash tank sodium hydroxide solution is added to neutralize the excess sulfuric acid. The salt formed breaks the emulsion and the water-salt layer can be removed. The oil may then be blended with other oils. The product is shipped in wooden barrels or lined steel containers. Tank cars may be used in shipping some fatliquor oils.

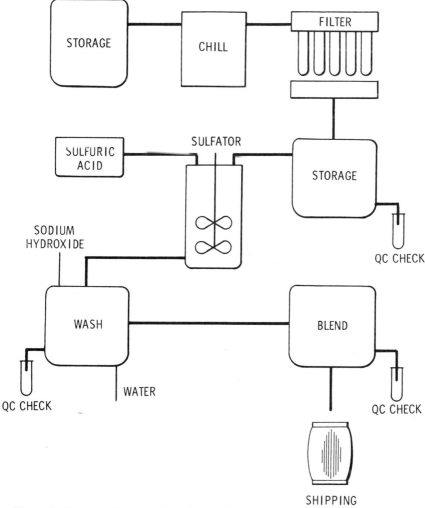

Figure 5   The manufacture of a sulfated oil.

Fatliquoring Emulsion Stability

A further understanding of the behavior of emulsions in fatliquoring can be obtained from a study of the emulsion of sulfated cod and mineral oil reported by Jean Poré in his fine book on fatliquoring, "La Nourriture Du Cuis, Méthodes et Principes."

Oils Used	Emulsion Obtained	Appearances	Hours Stability
Sulfated cod	Oil in water	Opalescent	40 hrs.
3% combined SO$_3$	Water in oil	Milky	48 hrs.
Sulfated cod	Oil in water	Milky	18 hrs.
1.5% SO$_3$	Water in oil	Milky	Unstable
Sulfated cod	Oil in water	Milky	24-48 hrs.
With 20% mineral oil	Water in oil	Milky	18 hrs. unstable

The formation of an oil in water emulsion or a water in oil emulsion will depend upon the amount of the oil and water components. A water in oil emulsion may be prepared, then added to a large quantity of water (i.e., in the drum during application) at which time a phase reversal may result. The resulting oil in water emulsion may have different stability characteristics than those formed by emulsifying the oil in the water prior to addition to the larger quantity of water. Even though the formulation may be the same, the addition methods may significantly alter the emulsion and consequently may alter the characteristics of the leather made.

Bisulfited oils

One of the most popular of the modern fatliquoring systems uses bisulfited oils. The bisulfited oils have application in the manufacture of chrome tanned leather. For lightweight, soft leather such as garment leathers or glove leathers, the penetration of the oil well into the leather is needed.

The reaction of the sodium bisulfite will take place at the double bond of an unsaturated fatty acid. The bond between the carbon and the sulfur of the bisulfite results in oils of broad stability to salt and pH variations.

Bisulfite additions:

$$R-CH=CH \quad + NaH\ SO_3 \longrightarrow$$

$$R-\underset{\underset{SO_3\ Na}{|}}{C}-CH_2$$

The addition of sulfur trioxide by the reaction of oleum (sulfuric acid contains sulfuric trioxide) will also result in carbon sulfur linkage and true sulfonates.

Sulfonation with SO_3:

$$R—CH_2—COOH + SO_3 \longrightarrow \underset{\underset{SO_3H}{|}}{R—CH—COOH}$$

With the bisulfited oils, the greater emulsion stability will allow greater penetration of the oil into the center of the leather. The development of bisulfited oils was a natural sequence following the increased demand for versatile oils for soft leather production. Since such oils command a premium price, their specific manufacturing details are held as proprietary information by the manufacturers.

Sulfonyl chloride

The sulfonyl chlorides are one class of fatliquors that have gained popularity because of their versatility. The synthesis takes place in several steps.

(1) A linear aliphatic hydrocarbon is made by the Fischer Tropsch process.

$$CO \, H_2—(CH_2—CH_2—CH_2) \, m + CO + H_2 \longrightarrow —CH_2—CH_2—CH_2——$$

(2) To this is reacted sulfur dioxide and chlorine.

$$——CH_2—CH_2—CH— + SO_2 + Cl_2 \longrightarrow$$

$$\underset{\underset{SO_2Cl}{|}}{CH_2—CH}—\underset{\underset{SO_2Cl}{|}}{CH_2—CH}— + HC \, l$$

The length of the hydrocarbon chain can be controlled as well the degree of chlorosulfonation.

(3) Upon neutralization with sodium hydroxide, the chloride hydrolizes and a sulfonated hydrocarbon results.

$$\underset{\underset{SO_2Cl}{|}}{CH_2—CH}—CH_2— + NaOH$$

$$\underset{\underset{SO_3Na}{|}}{CH_2—CH}—CH_2— + NaCl + H_2O$$

Figure 6 Sulfator for the manufacture of fat liquors. This particular sulfator is made of stainless steel and is equipped with heating coils, cooling coils, several agitation propellors, and an air supply. Through the use of the heating and cooling coils the temperature of the oil can be controlled during the sulfation or oxidation by the air bubbling. The propellors, located at the bottom with two on the sides, assure good stirring of the oil for proper reaction. This is a large piece of equipment, approximately 6 feet in diameter and 26 feet deep, and will accommodate approximately 20,000 pounds of oil. Also shown are external views at the top of the tank. The picture of the interior is taken through the manhole. *(Courtesy Atlas Refinery, Inc.)*

Because of the broad variation in the properties of these products, they are competitive with the bisulfited oils for soft leathers in the production of garment, glove, and soft shoe types.

Cationic emulsions

Both the soap and the sulfonic acid type emulsions are of the anionic type since they introduce negative charges to the micelle (droplet). It is also possible to use some of the organic nitrogen compounds known as the quarternary amines. These are positively charged compounds of the general formula below:

Cationic emulsion

$$R_1-\underset{\underset{R_3}{|}}{\overset{\overset{R_2}{|}}{N}}-R_4{}^+\,Cl^-$$

R_1—long chain amine

R_2, R_3, R_4—short chain amines

The action of the amines in oil and water emulsion is similar to that of the fatty acid in that the amines will form an oriented position at the interface between the oil and water and will result in a decrease in interfacial tension. The amines are positively charged and their acid anion counterpart remains in the water. The result, therefore, is that the emulsion is cationic (plus charge on the droplet) and is stabilized by acids. The peculiar nature of a cationic emulsion has some practical advantage in leather production in specialty cases.

Nonionic emulsions

Other materials, by virtue of their being semi-polar, may become oriented between the oil droplet and the water at the interface. Protein materials will be attracted to the oil because they are less polar than the water, yet they have a large number of carboxyl and amino groups and would have some affinity for water. The protein, therefore, would be oriented at the interface and result in a decrease in the interfacial tension. If the pH is kept near the isoelectric point, the proteins, particularly the albumins and globulins, will aid in dispersing the oil and in the formation of a fatliquor. These nonionic emulsions behave quite differently from the anionic emulsions of the sulfonated oils or the cationic emulsions of the quaternary amines. They have the property of forming relatively large oil droplets, and they are stable only in the pH range near the isoelectric point of the protein where the carboxyl groups and the amino groups are ionized.

In addition to protein materials, there are other hydrophilic (water-loving), semi-polar compounds which are nonionic in nature and aid in the formation of emulsions. In this classification are the polyhydric alcohols such as polyethylene glycol or methyl cellulose.

MECHANISM OF FATLIQUORING

In the treatment of leather with an oil, several phenomena are involved. It is the balance between these phenomena and the conditions of application of the oil, the nature of the oil, and the nature of the leather which determines the type and extent of lubrication of the leather fibers.

Practical fatliquoring

In a fatliquoring system using an anionic oil such as a soap or a sulfated oil, the oil droplet is introduced into the drum as an emulsion at a pH value where the emulsion is reasonably stable. Once it is in the drum, the oil droplets gradually begin to penetrate the leather. The pH in the leather is lower than is needed to maintain the stability of the oil, and the oil will be deposited on the leather. The introduction of the anionic oil changes the charge of the leather from plus to minus, and this change allows the next oil droplet to penetrate a bit further. The greater the degree of sulfation, not only the finer the dispersion but also the greater the penetration of the oil. Soap fatliquors have larger oil droplets and a greater tendency to be deposited on the surface of the leather than the same oil when dispersed by sulfation.

The attraction of the anionic oil group to the leather fiber brings about a stable bond upon drying, and the leather may have fixed oil. This will not be removed by a solvent, and when later extracted the leather may show some softness even though the fiber has been completely stripped of free oil.

Cationic oils have found limited application in the fatliquoring of most leather. Since the positive charge of the droplet does not permit much penetration of the fatliquor into the leather, the addition of a small quantity of the cationic oil at the end of the fatliquoring formulation will result in surface oiling by precipitation of the cationic fatliquor with the anionic fatliquors. This technique may aid in a decrease in grain crack or may result in a soft, silky feel to the leather without too much penetration of the oil into the skin.

Solvent fatliquoring

The principle of chemical bonding in the presence of an organic solvent is used in the so-called "solvent fatliquors." Solvent fatliquoring, as proposed by Von Fuchs, is based on introducing the oil into the leather in a high-boiling solvent. This solvent will displace the water in the leather, leaving a solution of the solvent and the lubricant. Upon slow evaporation of the high-boiling solvent, the lubricant remains absorbed on the fiber.

Another example of the same type of technique is used in the Secotan system. In the Secotan machine the leather is dehydrated by the forcing of acetone through from flesh to grain under pressure. Lubrication is accomplished in a second stage by a solution of acetone and oleic acid, or other lubricant, applied in a similar manner. The evaporation of the acetone leaves the oleic acid deposited in the leather, and a soft, supple leather with good fiber separation results. Modifications of these techniques are being used in practical fatliquoring application in compounded specialty fatliquors.

In recent years the term solvent fatliquoring has taken on another meaning. Many materials usable in lubricating leather cannot be directly emulsified by any of the common techniques. Some materials or fractions of leather lubricating materials must be solublized by the assistance of a solvent. Thus, a material may be dissolved in a mineral oil or hydrocarbon solvent and then this solvent emulsified by the use of other conventional fatliquors or emulsifying agents. With the increase in the availability of synthetic materials capable of lubricating leather and imparting desirable qualities the use of "solvent fatliquors" has increased. The practice is so widespread that it is no longer considered a specialty item by either the supplier or the tanner.

Stuffing

The treatment of leather with oils using a melt system rather than an emulsification system is known as stuffing. This is closely related to the old currying method and is one of the simplest forms of leather lubrication. The balance of oils, waxes, and greases imparts the desired characteristics to the leather; the stuffing greases are applied in a molten state while the leather is being tumbled in a heated drum. After the oils and greases have been distributed uniformly over the surface, the leather is removed from the drum and allowed to mull, once the initial heat has been removed by air cooling. In this system the natural fats used in the stuffing mixture contain some free fatty acids and other components which give them an anionic charge. These free fatty acids, anionic components, partially hydrolized fats, etc., are attracted to the leather fiber due to their polar nature. The other nonpolar components are then deposited on the leather fiber by physical absorption. As the lubricant is melted and mulled it penetrates further into the leather, and eventually the entire leather becomes lubricated, although in some instances a completely uniform distribution of the oil from grain to flesh may never be achieved.

A form of stuffing may also be through the use of sulfated animal oils and a large proportion of raw oils and petroleum derivatives. In this case, the stuffing mixture may be emulsified and applied somewhat like a fat-liquor. The liquor ratios are usually quite low and a much smaller quantity of water is used than in the usual fatliquoring system.

TYPES OF OIL

In addition to the method of application and the emulsion formation characteristics of oils, the ultimate lubrication properties and the physical characteristics of the resulting leather depend upon the types of oil used. Fats and oils used in leather manufacture fall into three broad classifications: (1) the triglycerides, (2) the fatty esters, and (3) the mineral oils. These simple classifications cover almost all the leather-lubricating materials. The most important are the triglycerides of the fatty acids. These include fish, plant, and animal oils. They are similar in that they are esters of glycerin and are different in that they are made up of different fatty acids. They may be characterized by their analysis of such factors as free fatty acid content, iodine value, melting point, titer, and saponification value. The natural fats contain almost no free fatty acids when they are obtained from the animal. Upon standing, and with bacterial action, some hydrolysis may take place with the formation of glycerin, and a free fatty acid may be broken from the fat. After further standing, additional bacterial action may take place which may cause this long chain fatty acid to break into two shorter chain fatty acids, due to bacterial oxidation at a double bond in the fat. This also results in an increase in the free fatty acids—a generally undesirable effect in that they contribute to the formation of spue and have little lubricant value. The unsaturated lower melting point fatty acids have value in some fatliquoring formulations.

Triglyceride of oleic acid (animal type fat)

$$H_2C-O-\overset{\overset{\textstyle O}{\|}}{C}-(CH_2)_7CH=CH-(CH_2)_7-CH_3$$

$$HC-O-\overset{\overset{\textstyle O}{\|}}{C}-(CH_2)_7CH=CH-(CH_2)_7-CH_3$$

$$H_2C-O-\overset{\overset{\textstyle O}{\|}}{C}-(CH_2)_7CH=CH(CH_2)_7-CH_3$$

Ester: cetyl laurate (sperm oil type fat)

$$CH_3-\left((CH)_2\right)_{10}-C\overset{O}{\underset{O-(CH_2)_{15}-CH_3}{}}$$

Iodine value

The degree of unsaturation of the fatty acids in a fat is measured by the iodine value, an analytical value whereby the unsaturated bonds of the oils are measured by the absorption of iodine. Iodine values encountered in the natural fats range from 10 (coconut oil) upward: palm oil 53, tallow 56, neatsfoot oil 85, soy bean oil 135, cod oil 150. The higher the iodine number, the greater the degree of unsaturation. High iodine values indicate low melting points and soft lubricating value for the oil. Very high iodine values, which indicate ability of the fats to polymerize, are obtained in some drying oils. In the case of leather, such drying oils may have distinct practical effects on the temper after it has dried completely. High iodine values also indicate the possibility that an oil will yellow eventually when applied to white leathers.

Melting point

Melting point, or cloud point, of an oil is a distinct indication of its quality and lubricating properties. High melting point oils (above 20° C) have a tendency to be firmer and have poorer lubricating qualities than those of low melting point.

Titer

Titer is not the same as melting point; rather, it is the melting point of the fatty acids obtainable by the chemical breakdown of the fat in question. It is useful, in an oil laboratory, to determine the characteristics of a particular shipment of oil and to identify oils. Titer is not generally used by tannery chemists.

Saponification value

In the mixed oil or the natural oil the triglyceride may be broken down to form soap and glycerin. Materials that cannot be broken down indicate the presence of unsaponifiable materials. The amount of unsaponifiable materials present may be a factor in the identification of the oil or may be an indication of an adulterant. The saponification value is a specific analytic value relating the amount of potassium hydroxide (milligrams potassium hydroxide per gram of oil) consumed in the saponification of a prescribed sample of oil. Since data is available on the characteristics of individual oils, this value may be used as a means of identification and also as a check on purity.

Cold test

The term "cold test" is an industry term and does not refer to a particular analytical procedure. A 20° F cold test oil may have a cloud point of approximately 26° F and a pour point of perhaps 10–15° F. The cloud test is more often used as an analytical procedure which involves the gradual cooling of the oil under a prescribed condition and observing the condition of the oil at 2° intervals on the even degree readings. The point at which a visible cloud appears is the cloud point. Since the method specifies observations at 2° intervals, the cloud point is always given in even numbers.

CHEMICAL NATURE OF THE NATURAL FATS

Since the natural fats which are of interest to us are primarily triglycerides, their nature is determined by the fatty acids occurring in them. Practically all naturally occurring fatty acids have an even number of carbon atoms. The shorter chain saturated fatty acids, C-6, C-8 and C-10, are found in coconut and palm oils, milk fat, and other soft oils. C-12 fatty acid, lauric acid, is found in sperm oil. The saturated fatty acids of the C-16 and C-18 category are common to the animal fats and many of the vegetable fats. The fatty acids in the C-24 and C-25 category are found in the hard natural waxes such as carnauba wax and beeswax. The unsaturated fatty acids, primarily of the C-18 type, are quite common in the animal and vegetable oils. Those fatty acids having more than one double bond are classified as drying oils, such as linseed oil and cottonseed oil. Some fatty acids may contain additional hydroxyl groups such as lanopalmic (C-16 hydroxy saturated) found in wool fat and ricinoleic (C-18 hydroxy unsaturated) found in castor oil. Both wool fat (lanolin) or wool grease and castor oil are common fatliquoring materials when sulfated. The makeup of the natural fats would be evident on consideration of a few specific cases that are typical of their classes.

Fatty acids

Caproic
  (C_6)
  saturated

$$CH_3-CH_2-CH_2-CH_2-CH_2-C{\overset{O}{\diagdown_{OH}}}$$

Caprylic
  (C_8)
  saturated

$$CH_3-CH_2-CH_2-CH_2-CH_2-CH_2-CH_2-C{\overset{O}{\diagdown_{OH}}}$$

Capric
  (C_{10})
  saturated

$$CH_3-(CH_2)_8-C{\overset{O}{\diagdown_{OH}}}$$

Lauric
(C_{12})
saturated

$$\mathrm{CH_3{-}(CH_2)_{10}{-}C}\begin{smallmatrix}=\mathrm{O}\\ -\mathrm{OH}\end{smallmatrix}$$

Palmitic
(C_{16})
saturated

$$\mathrm{CH_3{-}(CH_2)_{14}{-}C}\begin{smallmatrix}=\mathrm{O}\\ -\mathrm{OH}\end{smallmatrix}$$

Stearic
(C_{18})
saturated

$$\mathrm{CH_3{-}(CH_2)_{16}{-}C}\begin{smallmatrix}=\mathrm{O}\\ -\mathrm{OH}\end{smallmatrix}$$

Oleic
(C_{18})
unsaturated

$$\mathrm{CH_3{-}(CH_2)_7CH{=}CH(CH_2)_7{-}C}\begin{smallmatrix}=\mathrm{O}\\ -\mathrm{OH}\end{smallmatrix}$$

Ricinoleic
(C_{18})
hydroxy unsaturated

$$\mathrm{CH_3{-}(CH_2)_5}\underset{\mathrm{OH}}{\mathrm{CH}}\ \mathrm{CH_2CH{=}CH(CH_2)_7{-}C}\begin{smallmatrix}=\mathrm{O}\\ -\mathrm{OH}\end{smallmatrix}$$

Lanopalmitic
(C_{16})
hydroxy saturated

$$\mathrm{CH_3{-}(CH_2)_5{-}}\underset{\mathrm{OH}}{\mathrm{CH}}\mathrm{{-}(CH_2)_8{-}C}\begin{smallmatrix}=\mathrm{O}\\ -\mathrm{OH}\end{smallmatrix}$$

Carnaubic
(C_{24})
saturated

$$\mathrm{CH_3{-}(CH_2)_{22}{-}C}\begin{smallmatrix}=\mathrm{O}\\ -\mathrm{OH}\end{smallmatrix}$$

Cerotic
(C_{26})
saturated

$$\mathrm{CH_3{-}(CH_2)_{24}{-}C}\begin{smallmatrix}=\mathrm{O}\\ -\mathrm{OH}\end{smallmatrix}$$

In considering the characteristics of the individual oils, a word of caution should be interjected. Fatliquoring formulations supplied by fatliquor manufacturers to the American leather trade are blends of oils that have been balanced in their degree of sulfation and adjusted for pH characteristics and lubricating characteristics for a particular type of leather. They may be broadly classified by the manufacturer as being oil of a certain type. The guide given below is a general description of the characteristics of the oils and their properties but does not define the characteristics of any particular manufacturer's products.

Castor oil

Castor oil is triglyceride containing a large quantity of ricinoleic acid. The hydroxyl group of this oil attributes to the water solubility of the oil and it is easily sulfonated. Sulfated castor oil is a good lubricant for light leathers.

Cod liver oil

Cod liver oil, or cod oil, is obtained from the codfish and is characterized by a high degree of unsaturation which gives it some drying properties. It has an iodine value of about 150. It has the characteristics of imparting fullness and mellowness to the leather when sulfated and used in fatliquoring.

Neatsfoot oil

Neatsfoot oil is obtained from the rendering of the hooves and feet of cattle. It is closely related to the body fats of cattle and rendered into neatsfoot type greases. These "stocks" or greases are then cold pressed into low cold test or commercial grades of neatsfoot oil. Raw neatsfoot oil, sulfonated and bisulfite neatsfoot oil are the most common types of oils used in fatliquoring. These oils are fractions of oils obtained from the rendering of beef body fats. Pressing techniques and in some cases solvent extraction is used to obtain oils of desired characteristics.

Moellon

Natural moellon is produced by the oxidation of raw cod oil in the tanning of sheep and goat chamois skins. Synthetic moellon is produced by the controlled aeration of raw cod liver oil to the desired oxidation fatty acid value; it results in more hydrophilic properties.

Sperm oil

Sperm whale oil is also a soft lubricating oil and contains spermaceti wax, which is an ester rather than a true oil. Sperm oil is rich in fatty alcohols, and upon sulfonation sperm oil becomes a very strong emulsifier. Therefore, it contributes greatly to the penetration of the fatliquor into the leather. The sale of whale oil is illegal in the United States. To fill the need, synthetic oils have been compounded by the fatliquoring manufacturers which have the leather making properties of sperm, but, may be unrelated in chemical structure to sperm oil.

Mineral oil

Mineral oils may be employed in fatliquoring, but they are undesirable in that they contain no polar groups and, therefore, do not have the power of being fixed by the leather fiber. The incorporation of mineral oils into the formulations may aid in penetration of the oils and better fiber lubrication. The objection to mineral oil is that since it is not bound, it may migrate, particularly after contact with water, resulting in a change in characteristics of the leather.

PRACTICAL FATLIQUORING

In looking over the available formulations of fatliquors used in different types of leather, one is impressed with the wide variety of leathers which can be lubricated with relatively little change in the types of oils being used. The degree of sulfation and the amount of oils that are used in applying the fatliquor varies considerably. There are, however, several general practices in the industry which are quite common from one tannery to another.

In the manufacture of light leathers more oil is carried on a unit weight basis than in the manufacture of heavy leathers. It is common practice in making suede leathers to treat the leather with fatliquor shortly after a light vegetable re-tannage and leave the leather in the crusted state without dye. The dyeing of suede leathers is accomplished in a later stage. In calf leathers the fatliquoring is usually part of the dyeing operation, as it is with side leather. In the manufacture of side leather the total quantity of oil needed varies with the degree of softness and the type of garment being made. Leathers of light weights in the 1½–2 ounce class will require more oil on a unit weight basis (15–25%) than the heavy leathers (6–12%), where more firmness is desired. If the leather contains pigments or clay fillers, as are used in white leathers, additional oil will be necessary to provide sufficient lubrication since much of the oil will be absorbed by the pigment. The lubrication of heavy leathers involves more commonly stuffing techniques employing tallow wool greases and perhaps some mineral oils. The quantity of oils used in these firm vegetable tanned leathers is low, in the neighborhood of 3–4%, and a high portion of raw oils is used.

TECHNIQUES OF STUDYING FATLIQUORING

In scientific studies that have been made of fatliquoring, two general techniques have been employed to determine the location of the oil in the leather: chemical analysis and histological investigation.

Under chemical analysis, the technique is to divide the leather by stratigraphic slicing into layers, from the grain through to the flesh, and to run oil analyses on individual layers. From these data, curves can be obtained which indicate the penetration of the oil as a function of the treatment. This is a tedious and time-consuming procedure except in well equipped laboratories and is not generally employed by tanners in the development of fatliquoring formulations.

With histological techniques it is possible to stain selectively the oil in the leather. In this technique the leather is cut on a microtome and placed in a bath containing the stain. The stain is absorbed by the fat, and after proper rinsing and viewing under the microscope, penetration of the oil can be observed. Both the sulfated oil and the neutral oils can be observed by proper selection of the stains. Investigations of fatliquoring and the physical properties of the leathers have indicated the relationship between the amount of oil used, the distribution of oils, and the physical properties of the leather.

The oil that is in the leather in fatliquoring can be classified as fixed oil and free oil. Free oil is easily extractible, whereas fixed oil is absorbed by the hide fiber and is not released by extraction.

LOOKING AHEAD

The methods used in the lubrication of leathers have undergone a significant change with the change in tanning materials. The development of chrome tanning brought about the introduction of the sulfated oils. In recent years the development of syntans and resin tanning agents has altered significantly the fibers to be fatliquored. The requirements of specialty finishes, water-resistant treatments, dry cleanable materials, and new shoe manufacturing techniques have also set new standards of leather lubrication and performance. As leather is constantly being changed in its requirements and its method of manufacture, there are becoming available many new classes of compounds which could have value as leather lubricants. Although the supply of natural fats should remain high, since they are for the most part by-products of the food industry, the new synthetic specialty lubricants should gain in popularity because of their wider range of versatility of properties which can be put into the leather. Since these are to be applied in conventional equipment, and probably out of water systems, they will probably be carried in organic solvents as emulsions.

REFERENCES

Battles, M. H., "The Chemistry and Technology of Leather" (O'Flaherty, Roddy, and Lollar, ed.), Vol. 3, p. 73, New York, Reinhold, 1962.

Becher, P., "Emulsions Theory and Practice," New York, Reinhold, 1965.

Bull, H. B., "Biochemistry of the Lipids," New York, John Wiley & Sons, 1937.

Gustavson, K. H., "The Chemistry of the Tanning Processes," New York, Academic Press, 1956.

Harrison, J. J., "Interpretation of Analysis for the Layman," Nopoco Chemical Company, 1947.

"Hide and Leather and Shoes Encyclopedia of the Shoe and Leather Industry," Chicago, Hide and Leather Publishing Co., 1941.

Hilditch, T. P., "The Chemical Constitution of Natural Fats," New York, John Wiley & Sons, 1941.

J. Am. Leather Chemists' Assoc., Supplement No. 9, Symposium on Fatliquoring (1962).

McLaughlin, G. D., and Theis, E. R., "The Chemistry of Leather Manufacture," ACS Monograph 101, New York, Reinhold, 1945.

Poré, J., "La Nourriture Du Cuis, Methodes et Principes," Société des Publications "Le Cuis," Paris, 1974.

"Processing Chemicals for the Leather Industry," Newark, Atlas Refinery Inc., 1966.

Retzsch, C. E., "The Chemistry and Technology of Leather" (O'Flaherty, Roddy, and Lollar, ed.), Vol. 3, 1962.

Stiasny, E., "Gerbereichemie," Steinkopff, 1931.

THE DRYING OF LEATHER

13

After coloring and fatliquoring, leather is ready for drying. It has been tanned, and the materials for tanning, coloring, and lubrication have been placed in intimate contact with the fibers. Some of these materials are still in the solutions between the fibers, and the reaction with the fiber is not in all cases completed. Drying is more than the simple removal of the moisture to bring the leather to a practical, usable form; it also contributes to the chemical reactions of leather-making. The drying of leather is one of the most important steps in practical leather quality. In modern leather manufacture, with the need for high-speed production and with the use of more exotic materials, drying and the nature of the methods used have become of great importance.

EFFECTS OF DRYING

In Wilson's early books on leather technology the drying of a gelatin cube was used as an example of drying leather. This is an excellent example, and full understanding of the phenomena involved is of great help to the leather technologist. Hide protein is a complicated interwoven network of natural protein materials having numerous reactive groups of both acid and alkaline. Hide protein has associated with it a large amount of water and is, in fact, a hydrophilic colloid. Wilson's example of a gelatin block is, in truth, a hide protein without its fibrous structure. On drying the gelatin block, moisture is removed from the outer surfaces of the gel and there is gradual migration of the moisture from inside the block towards the surface. Under very slow drying conditions, evaporation from the surface proceeds at a slow enough rate for the water being removed from the surface to be replaced by that migrating from the inside. With high-speed evaporation, however, the water from the inside cannot migrate rapidly enough, and the surfaces become dehydrated. This dehydration permits the gelatin molecules to attract one another and to undergo hydrogen bonding, or a cross linkage from one fiber to another. The outer surface becomes a quite different material from the inside of the block; it becomes a hard mass which will decrease the passage of water to the surface. In the case of leather, the same phenomenon can take place: under excessive heating, the outer

222

surfaces become hard and the inner parts of the skin remain moist with the entrapped water.

Properly tanned leather behaves much differently from the gelatin block. Tanning eliminates some of the hydrophilic groups on the protein and makes them no longer available for the absorption of water. Thus the water in the skin can migrate to the surface much more easily and evaporation can take place. Leather will dry more quickly and easily than the hydrophilic fibers of a raw skin.

One of the criteria of a proper tannage is whether or not the leather will dry soft. This will vary from one tanning material to another. Synthetic tannins (exchange syntans) and aldehyde tannage (particularly gluteraldehyde) will give soft leathers upon drying. The vegetable tannins may dry more firmly but can be wet easily. Leather, therefore, that is tanned by chemical reaction with a hydrophobic material will usually dry soft and easily, whereas a leather tanned with a hydrophilic material will dry much more slowly and may be more firm.

In vegetable tannage, although the fiber of the leather itself may be hydrophobic and the reactive groups on the protein tied up with tanning materials, the vegetable tannins themselves are hydrophilic molecules which, consequently, harden on drying. It is necessary that evaporation from vegetable tanned leather proceed rather slowly so that the water in the skin can migrate to the surface without causing case hardening.

During the slow evaporation of water from a skin containing vegetable tanning materials, or any material which can react with the hide protein, several reactions take place. First, since water is being lost, there is an increase in concentration of the soluble materials in the skin. If evaporation proceeds rapidly, these materials may be carried to the surface and deposited during evaporation. A slow evaporation, or a vaporization of the water from the fibers within the skin, will allow these materials to be deposited deeper in the skin, and they will not migrate as readily. The rate of evaporation is of great importance in leather quality because it may affect the deposition of tanning materials and oils.

The attraction of the leather fibers for one another, which exists in almost any leather, will result in some stiffness upon drying. In causing the physical dimensions to be decreased, the fiber attraction will also result in some physical shrinkage of the leather. Drying methods that involve mechanically holding the leather in an extended position will result in a larger area yield and may prevent leather curling from the uneven attractive forces within the fibers. Practical drying methods are designed to stretch leather to obtain a good area yield by preventing shrinkage on drying. Tacking, pasting, toggling, and vacuum drying all employ this principle.

It is customary to dry the leather to very low moisture content so as to bring about permanent fixation of the materials within it. Once dried, leather may be wet back but will not attain the condition it had before the initial drying. The tanning, dyeing, and fatliquoring reactions are all completed by the drying process. This effectively sets the fiber in its physical form. Drying, therefore, is a chemical as well as a physical activity and is of great importance to the practical quality of leather.

PHENOMENA OF DRYING

The earth's atmosphere can be considered as a gaseous solution of water dissolved in air. The solubility of water vapor in air is a function of temperature. At low temperatures very little water vapor is soluble in air, but at higher temperatures the amount is considerably higher. Figure 1 is a diagrammatic representation of much of the action associated with drying

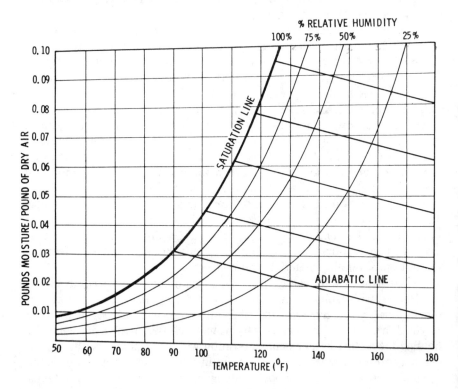

Figure 1 The moisture content of air at various temperatures. This chart is a simplified form of similar charts used in air conditioning, heating, and meteorological work.

and is of great practical use in solving drying problems. In spite of the large number of lines and the apparent complexity, it is really a very simple diagram from which practical information can be obtained.

On the vertical axis of the chart the units are pounds of water per pound of air. Along the base axis the temperature is in ° F. The saturation line indicates that air at low temperatures has very little water-vapor solubility. In other words, air will carry little moisture at the low temperatures, and the amount of water that is carried in saturated air increases significantly as the temperature rises. There is a series of proportionate lines to this indicating the moisture content of the air at various relative humidities.

For example, air at 55° F and 100% relative humidity, i.e., air saturated, will contain .01 pound of water per pound of air; at 90° F and 75% relative humidity, air will contain .024 pound of water per pound of air. From these two points we can see that in the summer air may contain considerably larger quantities of water than even saturated air in winter. If the air saturated with moisture in winter were drawn into a building and heated to 80° F the moisture content would still be .01 pound per pound, but the relative humidity would be only 40%. The air is dry and would easily remove moisture from leather.

Conversely, if air at 80° F and 50% relative humidity were cooled, the moisture content of the air would remain constant. The relative humidity, however, would increase with decreasing temperature, and at 60° F the air would be saturated. At this temperature the moisture in the air would condense on any object having a temperature lower than that of the air. This temperature is known as the "dew" point.

Let us consider what happens if we place a drop of water in contact with air that is not saturated. The moist air passing over the drop of water will remove some water from it. When this happens heat must be taken up from the air in order to satisfy the latent heat of evaporation of the water removed from the droplet. This loss of heat will result in a lowering of the temperature. This is thermally isolated, i.e., no additional heat has been added to the air or to the water droplet, but both the droplet and the air have been reduced in temperature because heat necessary to change the water from a liquid to a vapor form has been removed. Such a process, based on a change in temperature of a thermally isolated unit, is called an adiabatic change.

The amount of the temperature reduction and of the moisture increase has a constant relationship; consequently, in Figure 1 the adiabatic line followed in this phenomenon has a definite slope upward and to the left. If we extend this line to the point of saturation, the point of intercept is on the saturation line. This temperature is the "wet bulb" temperature. All atmospheres that are described as being along the particular adiabatic line in question have the same wet bulb temperature. From the measurement of wet bulb temperature and dry bulb temperature the relative humidity and the moisture content of the air can be obtained. This information can be used to calculate the amount of moisture that can be removed by the air.

For example, let us consider that we are operating a dryer that takes air from outside, heats it to 150° F, and passes it over the material to be dried; the exhaust from the dryer comes out as saturated air. If the air has a temperature outside of 50° F and 100% relative humidity (moisture content .008), the path followed would be horizontal at constant moisture content to 150° F on heating. The adiabatic cooling of the air during drying, with the picking up of moisture, would follow the adiabatic slope. The resulting saturated air (85° F) contains .026 pound of moisture per pound of air, and the dryer has removed .018 (.26–.008) pound of moisture from the material per pound of air heated.

Operating the same dryer in the summer, the air entering may contain 75% relative humidity at 90°F, or .023 pound of water. Heating the air to 150° F followed by adiabatic cooling during drying to the corresponding saturation point (wet bulb temperature) would increase the moisture content of the air to .038 pound per pound of air at 96° F. The moisture pickup in this case is .015 pound per pound of air.

It is evident that a dryer operating under apparently the same conditions in the summer and the winter, i.e., at the same temperature, will have different exhaust temperatures (85° vs 96°) and will remove quite different quantities of moisture (.018 pound vs .015 pound) from the material being dried. In fact, in the winter the same dryer will dry much more effectively than in the summer. If the dryer were operated with the same exhaust temperatures in both winter and summer, the differences in the rate of drying would be very evident. There may be differences in the drying rate when operating under these two conditions, and there may be a pronounced difference in the quality of the leather produced.

DRYING OF SOLID MATERIALS

In drying a solid material, the migration of water from the inside of the solid to the surface becomes an important factor. The nature of the material being dried is also of significance in that, as a dry state is approached, the equilibrium condition will depend upon the affinity of the entrained water for the material. The drying rate, therefore, will follow a distinct curve. In the initial stages the surface of the solid is saturated and the water will therefore evaporate at a constant rate. After drying has proceeded far enough so that the surface is less moist, migration of the moisture from the inside of the skin to the surface becomes significant. As the leather becomes dryer the rate of drying is related to the rate of migration of moisture from the inside of the skin to the surface.

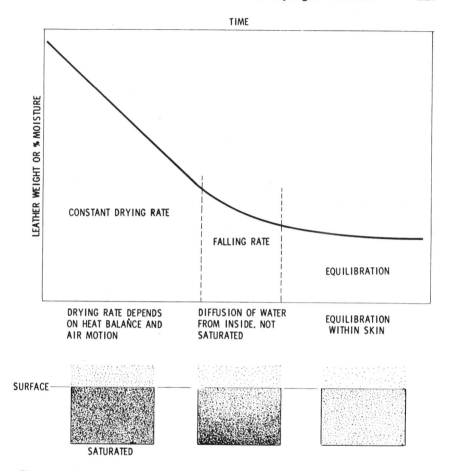

Figure 2 When leather contains free liquid water the rate of drying is dependent on the temperature, humidity, and motion of the air. At intermediate moisture content the rate of drying is dependent on the rate of diffusion of water from the inside of the leather. In the last stage of drying, the leather establishes an equilibrium with air.

In the final stages the rate of evaporation is determined by the release of moisture from the individual hydrophilic fibers. There will be a leveling off in the moisture content of the leather, and an equilibrium will be established. The rate of drying and the temperature and humidity of the air all play a part in determining the temperature of the leather during drying. The temperature of the leather will be of importance in obtaining consistent quality due to the chemical changes taking place in it during drying. In the initial drying rate period the evaporation of moisture from the skin can take place very rapidly, and the leather will approach the wet bulb temperature of the air. The temperature of the leather, then, in the initial stages of drying, will be considerably lower than the temperature of the air impinging upon it. The air closest to the surface will become saturated and will attain the true wet bulb temperature. As the leather becomes dryer the temperature of its surface increases because water is no longer evaporating fast enough to bring about as extensive adiabatic cooling. The leather temperature becomes higher, and overheating and case hardening may result. At equilibrium the air going into the dryer and the air coming out will attain the same temperature and adiabatic cooling is no longer a factor. Practical leather drying, therefore, becomes a matter of properly adjusting the temperature of the air in accordance with the moisture content of the leather at each stage of drying.

MOISTURE CONTENT OF LEATHER

Leather, being a hydrophilic material, will have a characteristic moisture content in accordance with its equilibrium with the air around it. The curve has a significant S-shape for all types of leather, i.e., the moisture content of the leather will be very low at very low relative humidities and will increase as the relative humidity increases. With a gradual increase in humidity, the moisture content of the leather will level off at a fairly constant level until the humidity of the air reaches approximately 80% relative humidity. At 80% and above, additional moisture will be taken up by the leather. These curves vary in shape, depending upon the type of leather being produced. The characteristics of the leather vary according to the range of this curve. At the low moisture content the leather is stiff,

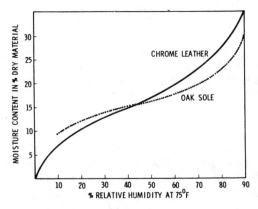

Figure 3 All leathers have some affinity for moisture vapor. In each leather a characteristic curve similar to those shown can be obtained. Although each leather is somewhat different, depending on the tannage, oils, etc., the moisture content will be less affected by changes in relative humidity under moderate conditions than by extremely high or low relative humidities.

cracky, and will shrink in size. As the relative humidity is increased, the moisture content, flexibility, and area of the leather increases. At the upper range the leather will be extended and will be more pliable. The normal characteristics of the leather are achieved near 50% relative humidity. For this reason physical tests are conducted on leather at specified humidities. Proper humidity control is necessary for the evaluation of leather for temper and for accurate measurement of the area.

For good shoe manufacturing practice, the cutting and lasting rooms should be maintained at constant relative humidity in order to aid in obtaining proper fit of the shoes and decrease the number of cripples (grain cracks) during the lasting operation.

The importance of maintaining proper humidity conditions in shoe manufacture cannot be overestimated. Crackiness in leather soles is a common ailment encountered in shoe factories. This complaint is predictable and will occur primarily in the months of January and February when the relative humidity of the air is very low.

PRACTICAL DRYING TECHNIQUES
Air drying

The earliest method of drying leather, and the one which is probably still in greatest use, is simple air drying. The skins are hung on hooks or sticks or

Figure 4   Air drying of light leathers. This simple drying system is very effective if climatic conditions permit. *(Courtesy Seta Leathers Ltd.)*

placed on horizontal racks. The leather is dried by the natural passage of air around it. If the conditions of temperature and humidity are such that the leather dries slowly, case hardening is avoided and good uniformity will result. The rate at which the air passes over the skins during air drying will determine the rate at which drying occurs. Overhead fans are usually used to circulate the air around the hanging leather.

To control drying, the air can be kept with the skins by shutting off the drying area. In this way the relative humidity around the skins will become higher and the drying rate will be reduced. In operating an air-drying system such as this, the temperature of the drying chamber, as well as the rate of release of moist air or rate of introduction of fresh air into the system, must be carefully controlled in order to obtain uniform drying conditions. Air drying has the advantages of (1) low capital investment, (2) no heat input, (3) little chance of case hardening, and (4) simplicity of operation. The drawbacks of the system are its low productivity and low area yield.

Tacking

A variation of the air-drying technique is tacking, where the leather is stretched out on boards and tacked. As the leather dries, the fibers tend to draw together and area loss results; this is avoided by tacking. The drawing together of the fibers during the tacking operation will tend to decrease the tendency of the leather to wrinkle and assume permanent folds. Tacking has the advantages of (1) low capital investment (but more than air drying), (2) no heat input, (3) no case hardening, (4) simplicity of operation, and (5) good area yield. The drawbacks are its low productivity (less than hanging) and higher labor costs.

Toggling

A toggling unit consists of a number of screens placed in a dryer having controlled temperature and humidity. The leather is stretched and held in place by a number of clamps (toggles) that hook into the screens. Toggling has the advantage of drying large quantities of leather in a relatively small space; in addition, it provides for stretching the leather during drying. Toggling has the disadvantages that constant temperature and humidity conditions are difficult to maintain and control.

Pasting

In a pasting unit leathers are pasted on large sheets of plate glass, porcelain, or metal with the grain surface to flat sheet. By this means the leather can be fully extended and the grain fibers so oriented as to give maximum smoothness and area yield; also, better grain characteristics can be obtained than with the air-drying or toggling techniques. The pasting plates are sent through a tunnel dryer consisting of a number (4–8) of zones of controlled temperature and humidity. The plates are suspended on an overhead conveyor and, with the leather pasted to them, travel through the dryer at a constant rate. When dry leather is stripped off, the plate is sent to a washer. The plates are washed and paste is applied automatically. Wet leather is applied to both sides of the plate and the cycle continues. By adjusting the temperatures and humidities in the various zones, the rate of evaporation from the leather can be controlled and overheating avoided. In the operation of pasting units, more severe conditions of temperature can be employed in the early stages of evaporation than in the later stages. Paste drying units are the most expensive single pieces of equipment in side leather tanneries. Paste drying is used on side, calf, kid, and split leathers.

Vacuum drying

The most significant advancement in the drying technique applied to leather is the development of the vacuum dryer. Vacuum drying originated in Europe and within the past decade has spread throughout the world; several hundred dryers are presently in use in the United States. In vacuum drying, the leather is spread out, grain down, on a smooth, usually a chrome-plated, polished steel surface. Heat is applied to this surface by a built-in head exchanger under the table. This temperature is maintained by thermostatic control of circulating hot water, and a hood is placed over the plate and then evacuated to aid in drying the leather.

In vacuum drying the same general rules apply to the heat balance as were discussed in connection with the air-drying technique. However, since the heat is being supplied directly from the plate to the leather, rather than being taken from the hot air, the entire drying system changes completely. Under high vacuum, the boiling point of the water in the leather is lowered, and with the cooling effects from the evaporation, the temperature of the boiling water is in the neighborhood of 100° F. At this low temperature fewer chemical changes take place in the leather, i.e., less hardening of the vegetable tannins and other case-hardening reactions. Since the heat is supplied directly to the skins, evaporation will take place in the fibers inside the leather, and the problems arising from migration of the tanning materials and oils are greatly decreased. Spreading out the skins on a steel plate gives some of the advantages of pasting in improving the grain and increases the area relative to that obtained by toggling. Vacuum drying techniques in general, however, give slightly less area yield than that obtained with pasting units.

The rate of drying will depend upon the thickness of the leather and the temperature in the platen. Chrome tanned leathers will stand higher temperatures than vegetable tanned; consequently, when drying chrome

Figure 5 Tacking of light leathers. Tacking aids in obtaining smooth leather with good area yield. *(Courtesy L. H. Hamel Leather Co.)*

Figure 6 Toggling of side leather. Some types of leather can be effectively dried in a toggling unit. By toggling, leather can be dried quickly using a minimum amount of space. *(Courtesy John J. Riley Leather Co.)*

Figure 7 The paste drying unit. The pictures illustrate paste drying—the application of side leather to a porcelain plate for pasting. The wet leather is placed on a plate which has already been scrubbed and has a light coat of paste. The leather is slicked out by hand and is applied on both sides of the plate. The plate is then carried by conveyor belt into the tunnel dryer and dried under carefully controlled conditions of temperature and humidity through a number of different stages. When dry, it is stripped from the plate and the plate is recycled through an automatic washing machine and paste spraying machine. The paste used is usually of the methylcellulose type, although some tanners prefer a starch-based paste. *Top: Courtesy Charles H. Stehling. Bottom: Courtesy Granite State Tanning.*

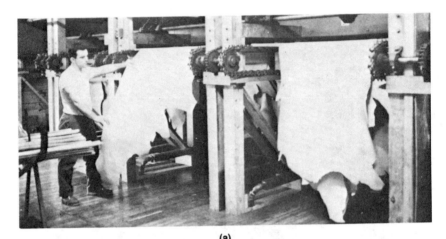

(a)

(b)

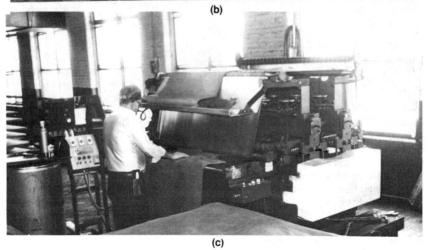

(c)

234

Figure 9 Several types of vacuum dryers are available to the leather industry. They vary in their design and in the relationship between the location of the plate and the vacuum cover. The plate is placed in a vacuum chamber, either by lowering the cover or sliding the plate into the chamber, and a vacuum is drawn. The moisture removed from the leather must be condensed in order to maintain high vacuum. A vacuum dryer, therefore, has with it a vacuum pump heater system and a cooling system. *(Courtesy Woburn Tanning Machine Co.)*

tanned leather, higher temperatures can be used in the platen. Vegetable tanned sides can be dried to approximately 14% moisture in ten minutes at a temperature of 125° F. Chrome tanned sides can be dried to the same moisture content in about three to four minutes at 180° F. Combination sides are dried in intermediate temperatures of 150°–175° F with drying time

Figure 8 **(a)** Conditioning for staking. Leather, after the initial drying (paste drying in this case) is re-wet and dried to a desired moisture content for staking. The drier used employs stick conveyors and temperature controlled tunnel drying. *(Courtesy Prime Tanning Co.)* **(b)** Staking machine. The purpose of staking is to knead the leather to soften it to the desired temper or flexibility after drying. There are two types of staking machines generally in use, the roll staker and the clamp staker. The roll staker works out the leather and makes it soft and mellow; the leather is worked through rollers through an extensive bend. The clamp staker is not as drastic as the roll staker and is more adaptable to heavier leathers, such as heavy upper leathers. With either machine, the operator considers the leather with respect to feel and moisture content and uses his judgment in staking certain areas of the skin to a greater extent than others. The moisture content is of great importance in staking in order to prevent cracking of the grain and breaking down of fiber structure. **(c)** Modern staking is done on automatic feedthrough machines. This machine carries the leather between two soft thin rubber belts. In the machine, the leather is staked by the vertical action of rounded steel rods. The machine is fast and accurate. *(Courtesy Granite State Leather Co.)*

Figure 10   Vacuum drying. Vacuum drying is popular for certain types of leather. This is the tanner's choice, depending upon the retannages used and the leather desired. The leather is spread on a smooth flat heated plate with the grain towards the smooth surface. The top cover is placed over the plate and the vacuum is applied. The leather is, therefore, drawn down flat onto the heated surface. Because of the vacuum employed, the drying is fast and at low temperature. This will permit using temperature-sensitive materials and rapid production. It is particularly popular in the manufacture of light leathers.

of approximately five minutes. The thicker the leather, the longer the time necessary to dry. Chrome calf skins, being considerably thinner (about 2-ounce leather), can be dried in as little as one minute at 172° F.

Since leathers are quite different, depending upon the tannage, fatliquors, and tanned effects desired, the tanner must adjust the temperature and time cycles of drying in accordance with the leather being dried.

The vacuum-drying technique has the advantage of being independent of weather conditions and can be used equally effectively anywhere in the world. The temperature and drying conditions are maintained by the equipment itself and are independent of outside influences. The timing cycle is controlled automatically by a timer. By the use of three or four units, a balance between labor and equipment can be attained to achieve continuous production relative to the efficiency of the labor involved. Vacuum-drying units are competitive in cost with paste dryers but are more expensive than toggling or other less sophisticated systems. The dryers are manufactured by several companies and their mechanical design varies. Each involves a heated platen with a movable cover with automatic timing cycles and with temperature and vacuum controls.

Figure 11   Crust sorting of shoe upper leather. After the leather is dried and staked "in the crust," it may be further sorted for temper and surface characteristics for selection into full grain, corrected grain, and embossed leather. The crust sort is in accordance with the needs of the customer. Most tanneries carry a large inventory in the crust that can be used for a wide variety of finishes. *(Courtesy Prime Tanning Co.)*

LOOKING AHEAD

Drying is one of the most important and most expensive operations in the manufacture of leather. Vacuum drying and the newer toggling machines give good area yield and desired leather properties.

Radio frequency is used in drying and for the initiation of chemical reactions in other industries. Radio frequency can be used to bring leather from a wet condition to less than 20% moisture in a few seconds without damage. It is energy efficient. Radio frequency may be used to a great extent in the future, but at present, it is used to a limited extent in finishing. High capital costs have delayed this development.

REFERENCES

Buch, L., "The Chemistry and Technology of Leather" (O'Flaherty, Roddy, and Lollar, ed.), Vol. 3, New York, Reinhold, 1962.
Pierson, F., *Das Leder,* **17**, 49 (1966).

FINISHING

14

The finishing of leather is probably one of the most complicated and least understood phases of the industry. Much of the technical writing that has been done in the past has concentrated primarily on the specific chemical problems of unhairing, tanning, coloring, and fatliquoring. The finishes, however, have been given scant attention. The reason for this is probably that the finishing of leather has long been more of an art than a science. The tanner has been receiving from his suppliers leather finishes of an increasingly specialized nature, and as the industry has become more complicated, he has come to have even less understanding of the finishing procedure. Finishing leather is not simply a matter of painting the surface to cover up the mistakes of the previous operations or to improve it by concealing scratches. Finishing contributes to the durability and beauty of leather and must be an integral part of the process. The compatibility of materials, tannage, coloring, and fatliquoring all play an important role in the character of the leather and the kind of finish it will take.

From a historical point of view, the leather industry was for many years a local affair, and the materials available for finishing were limited. As a consequence, the tanner put much hand effort into the development of beauty of the finish, and the cobbler simply worked with the leathers that were available and hoped to preseve the finish as best he could. In today's technology the point of communication between the shoe manufacturer and the tanner is less firmly established, and often the tanner does not completely understand the ultimate requirements of his product in its final use by the shoe manufacturer. The technology of manufacturing shoes has changed to such a degree that modern leather may be called upon to withstand strenuous conditions of temperature, mechanical operation, and chemical treatment even before the shoe is completed.

Modern finishes, when being applied in the tannery, require machine operations, such as pressing at high temperature, that cannot be duplicated on the leather once the shoe is completed. The tanner, therefore, must produce a complete finish of beauty and durability that is adaptable to modern shoe manufacturing methods; this must be done as simply as possible and with a minimum of labor. In the manufacture of shoes, redressing of the shoe is carefully avoided as much as possible by good shoe manufacturers.

˙The application of top dresses of an unknown nature on the finish will often cause more damage to the leather than the redressing process is worth.

REQUIREMENTS OF LEATHER FINISHES

The requirements of leather finishes are much more complicated than most other applications of coatings. The substrate differs from one type of leather to another, and the requirements with regard to flexibility, adhesion, and durability of the finish are very high.

Leather, prior to finishing, has been tanned, colored, fatliquored, and dried. At this stage the leather is resistant to bacteria and is flexible as is desired for its intended use. The leather, like wood, without finish lacks the feel, the texture, the resistance to scuffing and the resistance to water desired in the final products made from these remarkable materials.

In leather there are additional requirements. Leather is not a uniform material. It not only varies from hide to hide but from area to area within a single hide. The fiber structure variation from grain to flesh is also different in different areas of the hide. Each hide carries on its surface the history of the life of the animal. Scratches, insect bites, barbed wire cuts, brands, and even the effects of weather and the handling of the hide after flaying are marked prominently on the hide.

The uses of leather require uniformity of color, texture and physical durability. These requirements are not the same for each type of leather. Shoe leather uppers must stand repeated flexing and resist harsh weather. Upholstery must resist stains and the effects of continued bright sunlight. Other leathers all have their own requirements.

The lack of uniformity of the hides creates defects and finding normal structure is a major concern. Only a few hides in a particular lot will have a clear enough grain to be given only a light protective finish as ''full grain leather.'' For the remainder the finish must cover the blemishes and still maintain the beauty of the leather surface.

In leather finishing the finish system is a compromise between conflicting effects. If coverage of defects is the main problem some of the grain beauty will be lost. If resistance to scuffing is desired there is a danger of having a stiff varnished look. Each of these conflicting properties must be balanced in the final finish system.

Adhesion

Adhesion of a finished leather is dependent upon the condition of the leather, the type of tannage, and the kind of oils used in it. When the finish is applied it must stick; for this reason the leather surface must be ''wettable.'' The problem of adhesion involves not only the leather and the finish but also the various layers or coats of the finishing agents. A top finish need not adhere to the leather itself, but it must stick to the base coats of the finish to prevent peeling. Leather finishes require extreme flexibility and stretch for manufacture of shoes and for the flexing action of the shoe in use. Finishes that do not have good adhesion and good flexibility will peel and crack. The choice of resins and finishing materials is dependent to a great degree on the intended use of the leather.

Stability

Leather may be exposed to extreme heat during the manufacture of a shoe. When the shoe is worn in cold weather extremely low temperatures may be encountered. Thus the film must have a wide range of temperatures over which it is soft and pliable; it must also be hard enough to maintain the high gloss which is required.

Figure 1   The picture shows a feedthrough buffing unit, an example of automation in a side leather tannery. The leather is fed into the buffer, which is operated as a belt sanding machine and is equipped with brushes and a vacuum system to remove the dust. The side continues on through the machine, with the grain being sanded lightly, and then goes through an additional brushing machine to remove adhering dust. After the brushing a stacker is used to place the side leather on a high horse. A single pass is necessary on the buffing of the leather, and a full width cut is taken. In this series of machines working together, one operator can do the work formerly requiring three or four men. (*Courtesy The Aulson Tanning Machinery Company*)

   On the older models the width of the buffing paper wrapped on a cylinder could be as small as 12 inches wide, though 24 inches is more common. The machine is a feedback unit and the operator must take several cuts on the head area and several on the butt area to complete the buffing operation. Brushing was done either by a box brush in the feedback unit or by a feedthrough unit requiring two men.

Leather finishes, when applied, may contain some plasticizers or softening agents. In time these materials will migrate and be absorbed into the leather, resulting in hardening and embrittlement of the leather surface coating. The plasticizers can be replaced by the application of oils with the polish in some types of leathers. In others where the dressing is not applied, such as in upholstery leather, cracking may result. For good leather finishing, therefore, it is necessary to apply a binder of film-forming material which will remain soft and pliable over a long period of time.

The leather must be able to stand up to a reasonable amount of both wet and dry abrasion and to be refinishable with ordinary shoe polishing methods applied by the consumer.

Break of leather

For proper appearance and durability, leather must have a fine "break." This has been explained in other sections of the book. The application of finishes to leather will result in their penetration into the leather, which may affect the flexibility of the fibers near the surface and thus result in a change in the condition of break. If the finish remains on the surface and does not penetrate enough, the finish may tend to lift upon flexing and destroy much of the natural beauty of the leather. For some types of leather, particularly those that are buffed for pigment finishes, the filling action of the finish is of great importance. In this way uniformity of color and appearance can be obtained. In other types of leather, such as calf skin, where full grain appearance is desired, the filling action of the finish is limited and only a very light surface coating is possible. Leather finishes vary greatly depending upon the ultimate use of the leather and the type of surface desired. In any case, the conditions of flexibility, surface appearance, adhesion, and durability are essential.

COATING TECHNOLOGY

The coatings applied in leather technology may be divided into four broad classifications: (1) lacquer systems, (2) drying oil systems, (3) condensation systems, and (4) latex systems. Leather finishing, as applied in practical methods, may be the result of several of these coating systems. The systems differ from one another in their mechanism of film formation and in the chemical nature of the materials involved.

Lacquer systems

The formation of a film in a lacquer system is based on the evaporation of the solvent containing a film-forming material. A simple example of this is the dissolving of shellac in alcohol and applying this to a surface. Nitrocellulose dissolved in an organic solvent is also an example of this type of coating. This type of high-gloss lacquer is commonly used in household decorative enamels.

Drying oil system

The second basic group of film-forming materials is the drying oils. Into this classification fall the natural drying oils, such as linseed and tung, and also any material which will undergo polymerization upon drying. This is different from a lacquer in that the setting up of the film is not simply a deposition of a high molecular weight material; rather, it is a chemical reaction taking place between the dissolved film-forming materials and atmospheric oxygen. In the drying oils the film-forming material (a binder) is an organic chemical having a high degree of unsaturation. As the oil absorbs oxygen from the air, the unsaturated material is oxidized and reactive portions of the fatty acid molecule develop which can then polymerize with other fat molecules to form a continuous film on the surface.

Condensation systems

In condensation systems the formation of the film is due to a chemical reaction between the various components of the finish after application. The reaction may form a plastic or polymer in water between two molecules. Such systems are usually heat-activated and may be baked, glazed, or hot-pressed. Condensation or polymerization is used in the leather industry through protein-aldehyde reactions and with other resin systems. In this kind of finish the reactive components are usually mixed shortly before application, due to the limited pot life of the components.

Latex systems

In a latex system the binder is emulsified in water. When the latex is applied the water evaporates, or sinks into the leather, and eventually a phase inversion takes place. A continuous layer of the binder spreads on the surface of the material being coated and becomes a continuous film. Latex systems have their advantage in ease of application. The flammability and toxicity of organic solvents are largely avoided; the film can be diluted with water, enabling thin films to be applied.

Through modern techniques for forming copolymers, a wide variety of different lattices can be made which have a broad spectrum of properties. The versatility, low cost, and ease of application of such systems have led to their widespread use in the leather industry. The latex systems and dispersions of resins in an emulsion form are by far the largest and most widely used method of preparation in leather finishes today.

Copolymerization technique. Very seldom is it possible to polymerize a resin and obtain from a single polymer the exact characteristics desired in a coating resin. In order to obtain mixed polymers in a single resin, or copolymer, and also to obtain a stable emulsion for application purposes, the emulsion polymerization technique is used. Monomers are mixed together in the desired molecular ratios. This mass is then dispersed in water with a wetting agent, usually a soap or anionic detergent. A very fine particle size results, to which are added catalysts; the temperature is raised to bring about the polymerization within the micelles. Each micelle becomes a single particle of resins suspended in the water. When the temperature is lowered the particles of resin no longer have affinity for one another and are held in suspension.

Most resins are prepared with the particle size in the neighborhood of 0.2 micron and the total solids content of the emulsion system will be in the neighborhood of 40-50%. Emulsions are very stable and can stand both high and low temperatures without changing; the particle size will usually not change upon standing. These properties greatly broaden the methods of application of latex finishes.

A detailed study of the various polymers that can be used for copolymerizations in the formation of latex finishes for leather would be far too extensive to be included in a book of this type. We can only touch the high spots and consider the more common materials and their properties. Polymers used are substituted ethylenes, which upon polymerization will open up the carbon linkage in the chain. The nature of the components added to the chain will determine to a great extent the physical characteristics of the resin, its hardness, flexibility and stretch, etc.

Acrylic resins. A high degree of stretch from the acrylic acid esters of this type has led to their widespead use in leather. Methacrylic acid of the general form given is closely related to acrylic acid resin. It is harder than the acrylic acid resin and is blended or copolymerized to obtain the degree of flexibility and softness desired in the resin. One of the purposes of using a copolymerization technique is to build into the resins the characteristics, softness, and flexibility without additional plasticizers. By avoiding plasticization of the binder, the problem of instability of the resin in the finish over a period of time is greatly decreased.

The use of latex finishes has become dominant in the finishing of leather. The finishes are supplied as ''binders'' with resins of the desired characteristics and the color as pigments in suspension with little resin in the latex emulsion. The tannery finisher balances the resins and the pigments to obtain the desired result in each coat of finish.

Copolymers. Polyvinyl chloride and polyvinyl acetate are copolymerized to adjust the characteristics of the resin to the desired degree of stretch and

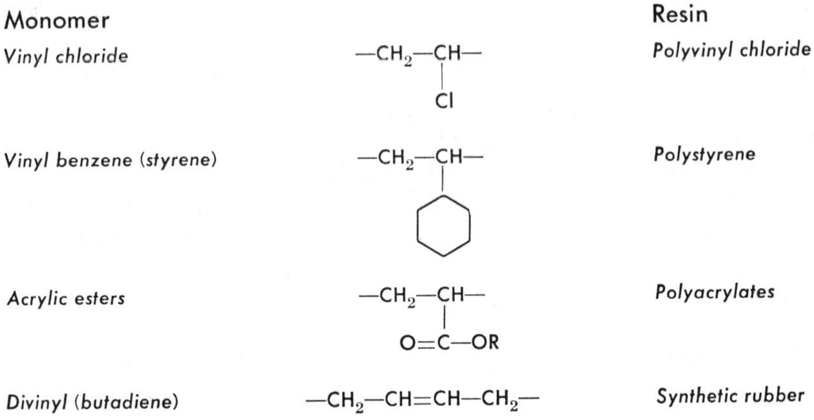

Monomer		Resin
Vinyl chloride	—CH$_2$—CH— $\|$ Cl	*Polyvinyl chloride*
Vinyl benzene (styrene)	—CH$_2$—CH—	*Polystyrene*
Acrylic esters	—CH$_2$—CH— $\|$ O=C—OR	*Polyacrylates*
Divinyl (butadiene)	—CH$_2$—CH=CH—CH$_2$—	*Synthetic rubber*

Figure 2　Substituted ethylenes.

hardness. Vinyl chloride films are tough and are used in top dressing. They are soluble in organic solvents and used in the same manner as nitrocellulose. Butadiene polymerizes to form a synthethic rubber—a soft, resinous material. Though it is far too soft to be used as a leather finish, it may be hardened by copolymerization with styrene, which alone would be far too hard and stiff to be used as a leather coating material. By proper balance of the monomers, a desirable copolymer can be obtained.

LEATHER FINISHING METHODS

Glazed aniline finishes

One of the oldest and yet most beautiful types of leather finishes is the glazed aniline finish. The method was developed early in the history of leather manufacture because of the readily available materials and the advantages that could be gained from handworking. Reproduction of glazed finishes by modern mechanized processes is very difficult, but even today large quantities of leather are being made with the process. The use of such finishes is confined to the more expensive types of leathers, such as reptile, kid, and high-grade calf. The base color for a leather to be made into glazed finishes is usually that obtained by the vegetable tanning process. The leather finisher then develops the final color by tinting the leather on the surface with a preliminary coat of an aniline dye. A basic dyestuff is generally used because of its strong affinity for the vegetable tanning material and because of its ability to develop shading of color on final glazing treatment. After application of the aniline dye, a protein is applied, in this case the globular

Properties and Applications of Leather Finish Systems.

	Physical properties								Area of use						Finish application				
	Flexibility	Adhesion	Dry abrasion	Wet abrasion	Cold crack	Surface feel	Break	Filling	Shoe upper leather	Soft upper leather	Upholstery leather	Patent leather	Calf	Kid	Base coat	Color coat	Spray coat	Topcoat	Impregnant
Non-aqueous																			
Vinyl lacquer	E	E	E	G	E	E	—	—	x	x	x				x	x	x	x	
Nitrocellulose lacquer	G	E	G	F	F	E	—	—	x	x						x	x	x	
Nitrocellulose emulsion	E	G	G	F	F	E	—	—	x	x			x				x	x	
Polyurethanes	E	E	E	E	E	G	E	E	x			x		x	x	x		x	x
Aqueous-system																			
Acrylic emulsions	E	E	E	E	E	G	G	E	x	x	x				x	x	x		x
Polyvinyl acetate	G	G	G	F	F	G	F	G	x	x					x	x			
Vinylidene chloride	E	F	E	E	G	E	G	G	x	x					x	x			
Styrene butadiene	E	G	F	F	E	F	F	F	x	x					x				
Polyurethane emulsion	E	E	G	G	E	G	E	E	x				x	x	x				x
Protein	E	E	G	F	F	E	E	E							x			x	

Grading

F—Fair
G—Good
E—Excellent

The above data is based on manufacturers' literature and the author's observations of common industry practice.

245

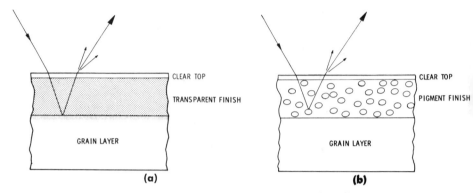

Figure 3 (a) Analine finished leather. The leather surface has been given a light coloring with a dye. The transparent finish and clear top allow light to be reflected from the grain surface of the leather to show the true beauty of the grain. (b) Pigment finished leather. In a pigment finished leather the light that is reflected is reflected from the pigment. The true grain surface of the leather, therefore, may not be seen.

type (casein or blood albumin). In addition to the protein, an ammonia dispersion of shellac is often employed.

The leather is coated with the protein material and allowed to dry. The dried leather is then placed on a glazing jack, a piece of machinery equipped with a solid cylindrical piece of glass, approximately two inches in diameter and six inches long. The machine is made in such a way as to allow fast stroking of the surface of the leather by a reciprocating arm. The pressure on the leather can be controlled and the surface of the leather glazed to the degree desired by control of the number of strokes in a particular area and by the pressure applied. By this system the operator can bring up good color tones and a high gloss finish in different areas of the skin as needed. The friction of the glass rubbing across the finish develops high temperatures which promote hardening of the protein and, at the same time, a quick smoothing action which results in a high gloss finish. The high temperature of the glazing jack also causes darkening of the colors and raising of some of the oils in the leather.

The characteristics of a glazed finish are high gloss, smooth surface, and subtle undertones of color. Since the finish itself involves no pigment, the porous fiber structure of the skin can be seen. It is possible to "look down into the leather" and perceive that the finish is part of it, providing depth and beauty.

The disadvantage of a finish of this type is the high cost of application. It may be necessary to apply as many as six different coats involving several glazings, platings, and staking and brushing operations to bring about the desired physical characteristics.

(a)   (b)

Figure 4   Glazing. **(a)** Glazing jack. The glazing jack shown is a small unit of the type used for reptile skins. Because of this use, it is called an alligator jack. The bed is approximately one foot long, and the glazing glass cylinder is about two inches in diameter. (*Courtesy The Aulson Tanning Machinery Co.*) **(b)** Glazing kid skins. (*Courtesy L. H. Hamel Leather Co.*)

Lacquer finishes

Lacquer finishes are formed by the deposition of high molecular weight substances from an organic solvent onto the surface of the leather. The earliest of these were the natural lacquers such as shellac; later nitrocellulose compounds were introduced. With the advance of polymer chemistry to give a wide variety of polymeric materials, the type and number of useful lacquers were greatly expanded. Lacquers have the advantages of high luster and gloss but have the disadvantage of poor covering ability and poor adhesion. Most finishes today are a combination of coatings, the lacquer being the outermost. Lacquer topcoats are compatible with the resin finishes underneath, and the advantages of the high gloss of the lacquer are obtained. The characteristics of a lacquer film are determined primarily by the properties of the material dissolved in the lacquer. The required characteristics of flexibility, stretch, and adhesion must be built into the resin system itself and cannot be obtained by simple plasticizing of hard materials.

The materials that make up a lacquer film have some residual reactivity, or at least have some absorptive capacity for one another, or there would be no real film-forming characteristics. The material laid down in the finish must have affinity for the leather fiber in order to obtain proper adhesion to the resin coatings of the undercoat. Glazing and high-temperature plating do not improve a lacquer-coated leather since the characteristic of a lacquer film is determined by the deposited resin itself and not by the physical adjustment of the film after it is laid down.

Latex finishes

With the increased mechanization of the leather industry, and with the desire to impart many different types of colors to leather and to obtain uniform quality at low cost, resin finish systems based on emulsions and pigments were developed. In a leather finish of this type the binder or film-forming material is in the form of an oil-in-water emulsion, along with the pigments and dispersing agents. The dispersion is applied directly to the leather by hand swabbing or brushing, which provides an even distribution of the emulsion. As the water soaks in, a coating is formed; the pigments and resins form a uniform film on the surface, as indicated in Figure 5. Two or three coats of pigment finish of this type may be applied, with less pigment in the later coats. Drying is usually done in a tunnel dryer, using infrared heat or forced hot air to evaporate the water from the surface.

After the application of two or more coats the leather may be plated to promote the fusion of the resin into the fibers and smooth the surface of the

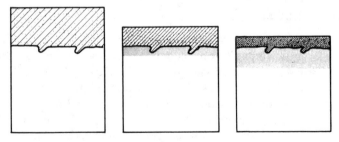

Figure 5 Finish applied to leather penetrates into the grain. This penetration causes a concentration of pigments and resin binder on or near the grain surface and aids in film formation.

leather. With the resin coat now in place, the topcoat is applied and better adhesion will result. The characteristics of the finish are determined primarily by the resin systems used in it. In the formulation of a finish of this type, the properties of the emulsions are also of great importance in determining the final finish.

Impregnation systems

The break of leather can be greatly improved by filling the spaces between the fiber bundles with a resin. This technique was first done by the impregnation of the leather with low molecular weight Urethane pre-polymers. The low molecular weight allowed penetration of the material in sufficient quantity and to the desired depth. The differences in the structure of the hide from back to belly area is partly compensated for by this treatment. The more open sections, which normally have a much poorer break, will take up more resin so the differences will be less evident.

The design of impregnation systems is based on the penetration of the desired amount of material in a limited time. The application and penetration time must be fast enough to work in the time allowed in the finishing machines.

The following points have been found to be significant in the design of impregnation systems to produce the desired results.

1. Harder polymers (having a higher modulus of elasticity) generally provide a better break than softer polymers.

2. The chemical nature of the polymer is important even if the physical nature of the polymers is similar.

3. The amount of impregnant is important. The break will improve with increased amounts of impregnant up to a point. Excessive amounts of impregnant will result in stiff leather.

4. The location of the polymer in the leather is of importance. The depth of the penetration and the amount of polymer in different locations will greatly affect the results.

In the development of a practical impregnation system, the balance of the resin and the control of the location of the resin is a matter of experience and artistry of the finishing expert.

Urethane finishes

The greatest recent advances in leather finishing are in the technology of the urethanes. Urethanes can have a wide range of properties. These differences in properties are built into the finish through a selection of the reacting materials and their ratios.

The essential building mechanism is isocyanate:

$$-N=C=O$$

Isocyanate is very reactive with available organic hydrogen as found in alcohols, amines, and acids. For example,

Isocyanate		Alcohol
R—N=C—O	+	HO—R$_1$

$$R-NH-C-OR$$
$$\overset{\|}{O}$$

If the isocyanate is difunctional and the alcohol is difunctional, a urethane will result.

$$O=C=N-R_2-N=C=O \quad + \quad HO-R_3OH-----$$

$$--C-N-R_2-H-C-OR_3- \quad - \quad C-N-R_2-N-C-O-R_3-O--$$
$$\quad \overset{\|}{O} \quad H \quad \quad H \quad \overset{\|}{O} \quad \quad \overset{\|}{O} \quad H \quad \quad H \quad \overset{\|}{O}$$

The size of the resin (molecular weight) will be controlled by the ratio of the two reactants. If the reactants are in equal balance and both reactants are difunctional, the theoretical molecular weight would be very large. With an excess of the alcohol, a stable resin of a desired molecular weight can result. This excess reactivity can be a problem since reaction with water or leather or any available organic hydrogen can result in desirable or undesirable side reactions. The system becomes very complicated. With the wide variety of isocyanates available and the reaction of amines and acids, the possibilities are tremendous.

Urethane technology is a specialty in itself and it is up to the finish manufacturer to build the desired properties into the system. The tanners are not specialists in resin systems and must rely on the finish supplier to meet his needs.

Some of the properties of urethane finishes have resulted in commercial successes that have come and gone with changes in fashion. The methods of attaining these properties are of interest and show the possibilities of other developments.

(1) Urethane lacquers. These have tougher, excellent low temperature properties. They have superior stretch and can be used for deep "wet look" leathers. These properties are permanent and don't change by drying out since they are internally plasticized.

(2) Quick dry system. This is a two package system; the first contains a

Figure 6   Automated smooth plating presses for side leather. The operator feeds the sides on a conveyor belt that passes through the jaws of two presses. The conveyor stops when the press is closed. The sides are completely pressed and will automatically be removed by the stacking machine. (*Courtesy The Aulson Tanning Machinery Co.*)

reactive group. This is not isocyanate, so it won't moisture cure. The prepolymer is mixed with its co-reactant, an isocyanate and solvent. The solution may react very fast upon application and be dry in about one minute. A thick bright single spray coat can result. It rarely needs dressing, and is the basis of an easy care system.

(3) Aqueous base urethane. With a proper balance of reactants, emulsion polymerization can be done with the urethane. This is very difficult due to the reactivity of the isocyanates with water. These systems are now available for aqueous base coats and top coats. Balanced urethane systems in aqueous emulsion can greatly enhance the natural look and feel of leather, while also adding an easy care advantage.

The development of impregnation systems and high-pigment finishes has resulted in greater utilitarian value of the leather and greater ease of cutting for the shoe manufacturer. On the other hand, the characteristics of beauty of grain and soft mellow feel associated with fine leathers are, in the opinion of many, greatly impaired. The increased "plastic look" in leather may, in the long run, harm the marketing position of leather in its competition with synthetic materials.

(a)

(b)

Patent leathers

Until recently patent leathers were made by a varnish-type system. Drying oils and plasticizers were applied and a thick, oxidized film of high-gloss finish was obtained. The introduction of urethane resins, with their flexibility and softness over a wide temperature range, has led to an increase in their use for high-gloss, opaque finishes of many colors. The urethane finishes have the advantages of easier application, higher-speed production, and greater durability than the older linseed oil process. Urethanes are also being widely accepted as top finishes on conventional leathers.

Moisture vapor permeability

One of the most desirable properties of leather is its ability to transmit moisture. This is an essential ingredient of comfortable shoe leather. The mechanism of this phenomenon has been studied by leather technologists in efforts to increase the removal of water from the inside of the shoe for cool comfort.

Several general statements may be made about this phenomenon.

(1) The leather fiber plays a part in the mechanism. A thin layer of water on the fibers allow evaporation at the leather surface.

(2) Even though the leather may be treated with water repellants such as silicones, the removal of water through the leather will continue as a vapor.

(3) The finish of the leather is important for shoe comfort since it can act as a vapor barrier.

Measurement of moisture vapor permeability is done by placing the leather between two differing known relative humidities. This is done by placing the leather as a jar lid with a screw cap. Then with a desiccant in the jar and a humid atmosphere outside, the weight gain rate can be measured.

Figure 7 Seasoning or finishing machine. (a) The purpose of this machine is to apply a uniform finish to the leather. It consists of two parts, one of which is a feed mechanism consisting of rollers and brushes whereby the finish is taken from a tub and distributed uniformly over the surface of the leather. The leather goes out onto a large, endless rubber belt where swabbing can be done either by hand or by automatic swabbing units. The finish is then allowed to dry by hanging on sticks as it passes through a drying tunnel. The use of the seasoning machine, the automatic swabbers, (b) the automatic stick hangers, and the take-off is another example of increased automation in a side leather tannery. (*Courtesy The Aulson Tanning Machinery Co.*)

The accompanying table shows the moisture vapor permeability of some polymer films.

	Transmission Rate $g/m^2/24$ hr/mil
Butadiene	680
Ethylene	4
Vinyl chloride	32
Methyl methacrylate	550
Cellulose acetate buterate	1500
Cellophane	1870

Sealed leather surfaces such as those with some thick vinyl chloride finishes have much less comfort than those with acrylic resin finishes with cellulose top dressing. The finishes with low moisture vapor permeability may not be comfortable for men's shoes or boots, but may have application for upholstery and other less critical comfort applications.

Water repellency

The most significant development in leather production in recent years is the development of water-repellent systems. New chemical treatments have contributed greatly to establishing modern standards of performance for leather in practically all categories. Water repellency can be obtained from several classes of compounds, such as alkenyl succinic acids, fluorinated acid-chromium complex, stearatochromic chloride complex, and silicones.

Alkenyl succinic acid is applied in a solvent system by utilizing the affinity of the polar end of the molecule for the leather. In this way the non-polar alkenyl end is exposed, greatly decreasing water penetration of the leather.

The fluorinated fatty acid-chromium complexes are used in textiles to aid in water and grease repellency. Similar properties are obtained in leather.

Some water repellency is obtained for garment suede by using stearatochromic chloride complex. The introduction of this product will also promote the dry cleaning ability of garment suedes.

The silicones are the most efficient means of increasing water repellency of any of the products available today. Silicones may be applied from a solvent system (either high-boiling hydrocarbon or chlorinated solvents) by surface treatment, or by a dip system. As a result of silicone treatment, shoes can be made that are for all practical purposes completely water-repellent. Performance equivalent to walking continuously in water all day without one drop of moisture seeping through the leather is common. These leathers still have the ability to transmit perspiration from the foot, and the comfort of the shoe is not impaired.

METHODS OF APPLICATION

Relationship of finish coats

The finishing of leather usually involves three types of coatings. In some cases the functions of these three coats may be combined into one or two finishing materials but the three functions are basic to any leather finish. The three coats are the base coat, the intermediate coats and the top coat.

Base coats. The function of the base coat is to tie the finish to the leather. The base coat may penetrate to a greater or lesser extent and may act as an impregnant. The primary function is adhesion. One of the most common leather finish problems in shoes and garments is poor finish adhesion. The leather may contain excessive oils or specialty chemicals that decrease the finish adhesion. The finish should be compatible with the leather.

The finish should not be too tackey. When applied the finish should dry fast enough to prevent the leather from sticking together and pulling the finish off when the leather is stacked.

The base coat and all of the leather finish coats must be flexible and have enough stretch as to avoid stiffness or cracking of the leather.

Intermediate coats. The intermediate coats must have the same desired properties of adhesion, flexibility and drying that are required in the base coat. The adhesion in this case is to the base coat rather than the leather. The intermediate coat must also be able to accept the top coat with good adhesion.

The intermediate coats are the decorative coats for corrected grain (buffed) leather. The pigment is mixed in the intermediate coats for the desired color and aesthetic effects. The intermediate, or pigment coat, may be the base coat if good adhesion can be attained with the system.

After applying the pigment coats, the dry leather may be plated. Plating can be done with a smooth plate or with an embossing plate. Smooth plating sets the finish to a smooth surface and in some cases sets the finish by heat curing. If desired, the leather may be plated with a patterned embossing plate to form a decorative grain surface. The finisher has the choice of many patterns to duplicate exotic leathers or reproduce the original grain of the leather.

Top coats. The top coat can be bright or dull. Any desired level of gloss can be attained by the use of inert additives. The main function of the top coat is to contribute ''toughness'' to the leather. The top coat also should contribute to the abrasion resistance, wet and dry crock resistance and chemical resistance.

Nitrocellulose lacquers were the main top coats for the industry for many years. In recent years more versatile materials have been used. The most common top finishes are as follows:

Nitrocellulose. The original top dressing now can be applied in a lacquer emulsion form. It is still popular in handbags and personal goods.

Polyvinal Chloride. This top dressing has good toughness and low temperature flexibility. It is commonly used in upholstery leathers.

Cellulose acetate buterate. This versatile top dressing has good resistance to yellowing from ultraviolet light. With the development of unit soles made of urethanes it was found that the cellulose acetate buterate finishes have good resistance to discoloration from urethane. This led to is wide acceptance in shoe upper leather.

Urethanes. These versatile materials have the best overall properties of any of the top dressings. They are expensive, however.

Hand finishing

The simplest method of finish application is by hand with swab or by brush. The finish, carried in a pan of water, is stirred and applied with a shearling swab to the surface of the leather. A second brushing with a mohair plush then follows, smoothing it out to an even coat. The finish is then air-dried. This method is particularly applicable to short production runs and experimental finishes. Hand-finishing methods are not desirable for high-speed production.

Seasoning machine

Continuous production is carried out by means of a finishing or seasoning machine in which the pigment finish, such as pigment latex finish, is carried in a tub, circulated by means of a pump into a pan, and then applied to the leather by a series of rollers and brushes as it passes through the machine. From the seasoning machine the finish goes to a rubber belt conveyor, or bolster, which lays the leather flat for swabbing to smooth the finish. Plush hand swabs can be used in a manner similar to hand finishing and may be mechanized with an automatic plush-swabbing device. The skins, after being swabbed, are hung on sticks and moved through a conveyorized drying tunnel.

Application by means of flow coater

The flow coater is particularly adaptable to methods of impregnation with solvent systems. In the flow-coating operation the finish is pumped into a tank above a conveyor belt. By means of a weir or orifice, an overflow of the finish falls in a continuous film into a tank below. The arrangement of conveyor belts is such that the leather can be passed through this "waterfall" of finish and be coated by it. Since the solids content of the finish is known, and the rate of flow is adjustable from the pump, the amount of finish falling in the curtain per unit of time can easily be calculated. Since the rate of travel of the leather through the curtain is known, the quantity of material deposited per unit area of leather can be determined. The quantity of finish can be adjusted to any desired amount in the flow coater. The falling curtain of finish does not touch the conveyor belt at any time, so the belt remains quite clean and the leather does not become stained on the flesh side. In the design of finishes for curtain coating, it is necessary that the viscosity and surface tension of the finish be adjusted so that proper penetration into the leather can be obtained. At the present time curtain coaters are used primarily for impregnation with urethane and acrylic finishes. The possibilities for the use of curtain coaters in other phases of leather manufacture appear to be very good.

Figure 8   Flow coating machine. (*Courtesy The John J. Riley Co.*)

Spray finishing

The application of finishes by means of a spray gun can be accomplished either by air sprayers or airless sprayers. Air sprayers operate on compressed air, but some finishes are best applied by airless sprayers which pump the finish to the nozzle of the spray gun. Finishes may be applied in spray booths on one piece at a time. For continuous large-scale production, automatic spraying equipment is employed. The leather is fed onto a conveyor belt consisting of a number of light lines. As the leather passes under a series of spray guns, the finish is applied evenly and effectively. The leather on the conveyor is carried through a drying tunnel where the drying may be accomplished by convection or by infrared heating. In the interest of economy, there are several devices that control the spray guns so that they spray only to the outline of the hides. The shape of the leather is determined by a series of sensing fingers, and this information is fed into a tape which acts as a memory device. By reading the tape at proper timing, control of the spray gun is adjusted so that just the outline of the side passing under the gun is sprayed. This technique results in considerable saving of finish and has been widely accepted by the industry.

Roll coating

Recent interest in air pollution, coupled by the increased costs of solvents has led to the search for methods of lacquer finish application which require less solvents. The use of solvent lacquer emulsions has been increasing since an even application can be achieved with less solvent.

One of the best methods of decreasing solvent use and conserving valuable finish materials is through the use of the roll coater. In the roll coater, a pan of finish is suspended above the work. A roller picks off the finish and transfers it by two or three rollers onto an applicator roll. The thickness of the finish on the applicator roll is closely controlled by use of a doctor blade. The applicator roll places the finish on the leather as it rolls over the leather. The clearance between the applicator roll and the back-up roll is adjusted so that good application is obtained but at no time does the applicator roll touch the back-up roll.

The system allows a very accurate application of a solvent based finish on the leather with the use of a minimum amount of solvent. Fast drying is achieved and costs are kept to a minimum. Unused finish is recycled in the machine with a minimum loss.

Rotogravure printing of patterns in leather is done in a similar roll press. In this case, color may be applied very accurately in patterns. The ink used is selectively carried by the printing roll in definite patterns. The patterns on the printing roll are made by the conventional methods of the printing

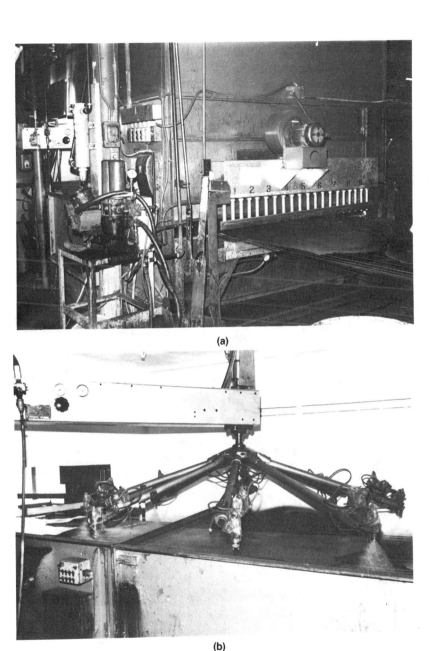

(a)

(b)

Figure 9 (a) The sides are laid flat and sent through the unit by conveyor. On the conveyor, there may be a sensing unit, as indicated here under numbers 1–10, which will read the shape of the hide, and control the spray gun, as it moves on down the line. (b) The spray guns, as a result of intelligence fed by the sensing units, will spray onto the hide, not on the empty conveyor. Some rotating spray heads also have their own sensing units, in the form of photo electric cells, mounted in the head itself. The system will aid in the saving of valuable finishes.

259

industry. The use of the rotogravure printer will allow a wide variety of patterns and even multicolor patterns on leather. As the techniques of printing are improved we can expect an increased use of this system in the industry.

Embossing

In order to obtain special grain effects, leather is often put through an embossing press with a characteristic grain pattern on the plate. Under conditions of high pressure and temperatures up to 250° F, the leather is shaped and a permanent pattern can be achieved. Smooth plating is used in most leather finish formulations. The high pressure and high temperature conditions give a smooth surface to the grain.

In addition to high-pressure embossing presses, lower-pressure hot plates

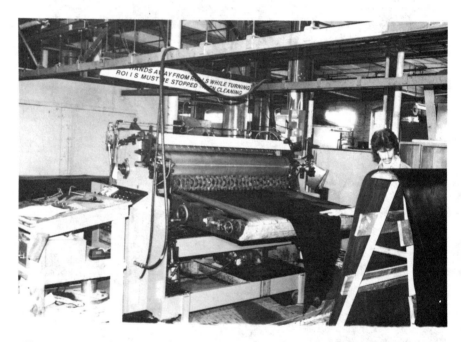

Figure 10 Roll coat finisher. With the high cost of finishes and also environmental regulations, the use of lacquers has been greatly aided by the application onto the leather by means of a roll coater. In this case the leather, as it is fed into the machine, is spread out by some rubber fingers. As the machine passes through, a roll on the top rolls finish directly onto the surface of the leather. The roll does not touch the conveyor. The finish is applied uniformly onto the roll from a feed source above, including spreading rolls. The quantity of solvent kept in such an application system is minimal, allowing for a saving in solvent quick drying and less discharge of volatile organic compounds to the atmosphere.

are used where the action of heat and a smooth surface are desired. The effect of hot, smooth plating may be both physical and chemical. The high temperature may cause softening of the resins, waxes, and oils of the finish and induce a change in the penetration of the finish and in the grain characteristics. Some finish materials are thermosetting, and the hot press may promote the film-forming reaction.

New presses have been developed based on the cylinder feedthrough type of operation. In these presses the weight is concentrated in a relatively small area; the two surfaces move at slightly different speeds and a high-temperature, high-pressure rub gives an effect similar to glazing. The primary purposes of the feedthrough press are, of course, the advantages of labor saving and greater production.

Boarding machine

Good grain effects can be obtained by rolling the leather upon itself and forcing the grain patterns into the leather. This is done by boarding. On a boarding machine the leather is placed between two rollers, rolled back upon itself, and then fed back out. By turning the leather approximately 60 degrees, a second directional boarding can be obtained; a third can be ob-

Figure 11 Embossing press. The embossing press is used not only to put a decorative embossed grain surface of the leather, but may also be used to smooth plate. Many finishes require a high temperature pressing to heat cure the finish. This is a modern hydraulic press. It has the advantage of faster production than is obtained in the mechanical presses that are normally found in the industry. The plates are heated by steam and kept at a desired temperature thermostatically.

tained by an additional 60-degree turn. The boarding operation, if used at all, usually occurs just before the final finishing coats.

SHOE SIDE UPPER LEATHER

Side upper leather is primarily the buffed type. The re-tanned crust leather is lightly buffed to remove surface imperfections and prepare the surface of the leather for the finish. This type of finish is usually pigmented, with two to three coats applied as a base and one or two as top coats. The leather may be impregnated, prior to finishing, with the urethane in a solvent system or with a resin emulsion (usually acrylic) which will penetrate into the leather and tighten the grain. After the initial impregnation the pigment finish is added (possibly containing some dye) in the aqueous phase of the finish system. Application of the finish is usually by the seasoning machine, applying an even coat and swabbing off. After one or two base coats of the pigment, the leather may be sent through a press to smooth the surface and prepare it for the top dress. The latter is applied with one or two coats of lacquer emulsion, usually by a spray using organic solvent or a water-dilutable type of lacquer emulsion. Many different application systems are used in side leather depending on the fashion effects desired.

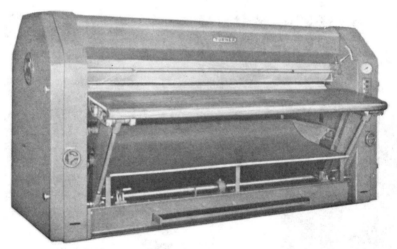

Figure 12 Boarding machine. The boarding and softening machine which has been in use in the leather industry for many years replaces the hand-boarding operation. This operation in the finishing room is to roll the leather so as to form grain wrinkles in a desired pattern. Leather may be boarded one, two, or three ways, each taking place at different angles. Boarding in one direction results in a willow effect; boarding in two directions gives a box effect. Boarding accentuates the natural grain pattern of the leather. (*Courtesy Turner Tanning Machinery Division of United Shoe Machinery*)

In finishing side leather, the cost of finish application is of great importance and very little hand work can profitably be employed. Glazing and brushing techniques are avoided in side leather production as much as possible.

CALF LEATHER

In the finishing of calfskins, the grain surface is of greatest importance. Calfskins are sold as full grain leathers, and as such, the pigment application on the calf leather is kept at a minimum. Finishes are more of the aniline type, the light dye coat being applied first. Many leathers are still finished by the application of dilute protein coats followed by glazing, ironing, and plating operations. These mechanized operations build up the smoothness and chemical resistance of the finish as desired. Top coats in this case may consist of waxes for final dressing.

KID LEATHERS

Kid leathers, similar to calf leathers, are finished to enhance the beauty of the grain. Therefore, finish methods similar to those used for calf leathers are employed.

Figure 13   Finish sorting. Each individual piece of leather, as it is completed and readied for shipment at the tannery, must be inspected, trimmed, and sorted by grade. Although all the leather in a particular pile may be going to the same customer, it must be graded for purposes of proper pricing. (*Courtesy Amoskeag Leather Finishing Co.*)

UPHOLSTERY LEATHERS

Upholstery leather is quite different from shoe upper leather both in its chemical characteristics and in its use. It is vegetable tanned and has sufficient oil to maintain flexibility over a long period of time without further dressing. Upholstery leather is exposed to a wide range of temperature, so the finish can neither become tacky when warm nor crack at low temperature. The early nitrocellulose finishes were replaced with resin dispersions, and these in turn have been displaced by solvent systems based on vinyl resins.

FINISHING SPLITS

Splits are an extreme case of a buffed leather in that they are taken from the flesh side of side leather and have a very open grain structure. A "synthetic grain" can be built up by applying finishes of high binder content and good filling action to the surface of the split; the addition of pigments will also help to build up a synthetic grain. By this method a leather approaching highly buffed side leather in appearance can be made. Regardless of the finishing procedure splits will never attain a value equal to that of grain leather. Consequently, the finishing must be done in very few coats with as little hand application as possible. The balance between costs and the added value in the finishing of splits is particularly delicate, and a very close watch must be made of the economics in the finishing system.

FINISHING FANCY LEATHERS AND REPTILE LEATHERS

In many fancy leathers, particularly reptile leathers, the fine grain surfaces of the leather are accentuated by use of the glazing process. The vegetable tanned reptile leathers may be given a light coating of albumen and glazed on the very small glazing jack. By use of the small glazing head the characteristics of the grain can be controlled from one small area to another, and the maximum value can be achieved from these very expensive skins. Finishing of reptile skins is an extreme case of the use of hand work to develop a high grade finish; this is opposed to the extensive use of chemicals employed in the low grade side leathers and split finishing.

ENVIRONMENTAL CONSIDERATIONS

One of the main problems of leather finishing is the environmental problem of organic solvents. One of the most effective application methods for impregnation of leather with resins is by solvent application. Lacquer finishes and top coats also are easily applied and are quick drying. Environmental regulations have all but prohibited direct application of solvent-based fin-

Figure 14 Automatic electronic measuring machines measure, mark, record data, and total the footages. (*Courtesy Amoskeag Leather Finishing Co.*)

ishes. The approach that has been developed is the use of lacquer emulsions. This presents the problem of drying both the water carrier and the solvent. The problem is now satisfactorily solved and the lacquer emulsions are in common acceptance.

LOOKING AHEAD

The area of leather finishes is one that is changing very rapidly with the availability of new resin systems from chemical manufacturers. The industry is being called upon to produce leathers of greater beauty and durability of finish than ever before. For a true "leather look," the aniline type of finish is desired, yet the public expects durability under the most adverse use conditions. Through the combined effort of both finish manufacturers and tanners, we can expect to see rapid changes in the standards of acceptable quality and durability of finishes in the very near future. This durability must be attained without losing any characteristics of beauty in grain and texture. If these qualities are sacrificed for uniformity and durability, leather will gradually approach the characteristics of substitute materials and may lose its reason for existence.

The finishing methods outlined in this chapter are based on present technology. In the coatings field for textile, paper, plastics and other surfaces a number of techniques are used which have been applied to a greater or less extent to leather finishing. These included silk screen printing, rotogravure printing, transfer coating techniques and lamination techniques. All of these will probably play a greater role in the leather finishing of the future to develop greater latitude in the aesthetic properties in the final product.

The coating technology employed in metal finishing, plastic technology and paper coating has resulted in the remarkable use of high energy radiation such as ultraviolet, electron beam, and radio frequency curing of coating materials. Leather is limited in some of these applications due to its protein base. The practical application of these new drying and curing methods will probably be the next big technical advance in leather finishing.

REFERENCES

Freudenberg, W., "International Dictionary of the Leather and Allied Trades," Berlin, Springer-Verlag, 1951.

Gould, R. F., "Plasticization and Plasticizer Processes," Washington, D.C., American Chemical Society Publications, 1965.

"Hide and Leather and Shoes Encyclopedia of the Shoe and Leather Industry," Chicago, Hide and Leather Publishing Co., 1941.

J. Am. Leather Chemists' Assoc., Supplement No. 10, Report of the Symposium on Surface Characteristics of Leather—June 19, 1963 (1965).

"Leather Chemicals," Philadelphia, Rohm and Haas Company, 1955.

Lewis, W. K., Squires, L., and Broughton, G., "Industrial Chemistry of Colloidal and Amorphous Materials," New York, Macmillan, 1942.

Miler, F. D., "Cellulose Nitrate," New York, Interscience Publishers, 1955.

Orthman, A. C., "Tanning Processes," Chicago, Hide and Leather Publishing Co., 1954.

Payne, H. F., "Organic Coating Technology," Vol. 2, New York, John Wiley & Sons, 1961.

Spiers, C. H., and Blockey, R., Jr., *B.I.O.S. Survey Report,* No. 27 (1950).

"25 Jahre Corialgrund," Ludwigshafen am Rhein, Badische Anilin and Soda Fabrik AG, 1956.

Wilson, J. A., "Modern Practice in Leather Manufacture," New York, Reinhold, 1941.

Wilson, J.A., "The Chemistry of Leather Manufacture," Vol. 2, New York, Reinhold, 1929.

FURS & FUR TANNING

15

Since before the beginning of recorded time, furs have been an important part of the life of man. Furs were the earliest type of clothing for primitive man. In tropical climates, furs are chosen for the beauty of their exotic color and texture. In arctic climates, furs are used for warmth. The types of tannages and preparation of the skins are historically a matter of using local materials with consideration for the ultimate use of the final product.

The primitive decorative furs in the tropics were dried, vegetable tanned, or possibly smoke tanned. In an arctic climate, the tannages were usually based on the effect of the oxidation of natural fat of the skin. The utilitarian furs of the arctic are made into parkas and boots which are quite satisfactory and will keep the wearer warm. Skins tanned in the arctic manner would have no value in a temperate or tropical climate since the bacterial attack would occur rather quickly.

In a consideration of the modern fur industry, we should limit ourselves to skins of true economic value. There has been some public concern that the fur industry is forcing the slaughter of many wild species throughout the world. As a result of concern for endangered species, some think that the furs should not be worn in order to preserve the wild animals. It is true that the search for particular types of skins, namely spotted cats and tigers, etc., has resulted in the decrease in numbers of animals and many animals are on the endangered species list. On the other hand, some furbearing animals are selectively bred and raised under closely controlled conditions of diet and cleanliness for the highly developed, sophisticated fur industry. We can, therefore, divide the available sources of raw materials for furs into two types of sources.

(1) *Wild Animals*. In this classification we would include all the wild animals used for furs such as wolf, some types of fox, natural mink, marmot, and jackal, muskrat, racoon, opposum, and others.

(2) *Domestic Animals*. These furs include animals that are raised primarily for their fur. Ranch mink and certain types of foxes which are bred in captivity for fur use are in the true sense domesticated animals.

The following animals are those of greatest importance as commercial furs. Other species such as the sea otter, the spotted cats, and bears are no longer significant in the commercial fur trade.

(1) *Beaver*. Wild. Once common in Europe and Asia, it is now almost exclusively a North American species. The population is stabilized and present production is about 250,000 skins per year.

(2) *Fox*. Available both as wild and ranch types. Wild fox has a red brown color. Ranch types are bred for desired colors. Silver and black fox are color phases of the red fox. Total production in North America is about 150,000 skins per year.

(3) *Rabbit*. The most numerous of the furbearing animals. They have been used for dressing and as lining furs. Rabbit is very inexpensive. The animals have been bred to a wide variety of textures and colors to look like more expensive furs.

(4) *Mink*. The most important (highest dollar volume) fur in the world commercial trade. Wild mink are available but ranch mink are by far the more important. Mink are bred for color, size, and other desirable features. Total production is about 20 million pelts per year, about half of which is North American and the bulk of the rest is Scandinavian.

(5) *Muskrat*. These are common wild animals that are very significant in the fur trade both in Europe and North America. Total production is between 50 and 100 million pelts per year.

(6) *Karakul*. This is the second most important commercial fur in the world. The newborn sheep are slaughtered at less than three days of age. Some skins are taken from unborn animals to yield broadtail. The total number of skins produced in Southwest Africa, Afghanistan, and Russia is about 2 million per year.

(7) *Seal*. There are two types of seal used in the fur trade: the fur seal and the hair seal. The fur seal has a thick silky coat used extensively for coats. The hair seal ranges from yellow brown to dark brown. Only the very young of the hair seal are used.

FACTORS NECESSARY FOR A VIABLE FUR INDUSTRY

A fur industry requires a supply of animals with sufficiently good quality fur to make the skins marketable. The animals must be hunted, trapped, and killed in a manner to eliminate most of the damage to the skin at the time of trapping or between the time the animals are killed and the skins are marketed. It is, therefore, necessary to have effective field preservation techniques known to the people who are doing the trapping. Wild furbearing animals are in relatively remote areas, and the preservation of the skins taken is usually done by traditional methods that have evolved over centuries.

The field-preserved skins are marketed through dealers or brokers. The dealer may buy from many different trappers and may find it necessary to travel great distances to accumulate the skins. He may further treat the skins to protect them and make them more presentable for marketing.

travel great distances to accumulate the skins. He may further treat the skins to protect them and make them more presentable for marketing.

A commercial lot consists of tens of thousands of skins that are then re-sorted and sold by the dealer to the manufacturer or a broker. The skins are further processed, tanned, and dyed to make usable articles of commerce. The tanned and dyed skins may be sold again prior to reaching the garment manufacturer. The garment manufacturer may or may not own the skins he is working into garments. The garments may be made on contract for the wholesale or retail outlet.

The fur business is by nature a very high risk industry. The trapper has a very uncertain catch and seasons of very low personal income. The dealer buying the skins from the trapper takes a risk that the skins might not be properly preserved or the proper grade to have appeal in the present market. The skins must be processed properly, i.e., dressed and possibly dyed. This is done with some degree of risk in the tannage as a small amount of hair damage (or hair slip) would be disastrous to the value of the skins. The dealer must grade the skins and put them into the proper lot for sale to the garment manufacturer, either before or after the dressing.

Quality skins, available in sufficient quantity, are carefully sought for, but in any mixed lot there are a number of skins off-color, off-size, and off-grade, to be sold at a lower price. In dressing the skins, there is always the danger of damage during tannage or coloring due to improper processing. The garment manufacturer buying the skins runs the risks that are normally associated with selling a luxury item at a very high price.

The fur industry, therefore, is one of considerable risk, both financial and technical, from the time of the slaughter of the animal to the final retailing of the garment. The high costs of fur garments, therefore, are related to the economic practicalities of grading, skin spoilage, and trends of fashion.

PROCESSING OF FURS

The processing of furs is an ancient art which precedes the development of the grain leather industry. The fur industry is considerably smaller than the leather industry. The tanning of furs has been kept a jealously guarded secret for many years. Most published references on fur processing date back at least forty to fifty years, and the methods described are actually quite dissimilar to those in common practice in the industry today.

Dressing and tanning of furs is a very risky business and one that even under the best of conditions may result in the loss of a few skins. The tanning skills required are unique to the product and it would be a mistake to expect that a good shoe leather tanner would be able to make good fur skins.

Three furs will be considered in this chapter: mink, karakul, and shearling. All are important in their own way and illustrate the variations in technology and economically different segments of the industry.

MINK

Mink, as mentioned earlier, is the largest dollar value segment of the fur industry. The largest category of mink skins is the ranch mink of United States and Canada. Mink ranches range from a few animals raised on a hobby basis to large operations of tens of thousands of animals. The animals are bred in late winter with each female giving birth to an average litter of four kits in the spring. The kits are separated from their mother at an age of about two months. At four months they must be separated to individual cages to prevent fighting. They are carefully fed with a high protein diet which is a major expense of the operation. When the pelts are mature (early winter), the animals are pelted. The skins are removed without slitting up the belly. This will preserve the fur better without cutting in valuable areas. The skins are stretched on flats and dried. The skins are sorted and matched with both male and female in proportion to make a coat in a single lot. The furs are usually sold at auction either dried or in the tanned condition.

Tanning and dressing mink

The processes outlined for mink are similar to those used for muskrat, opossum, and other small animals. The skins are wet back with particular attention to rehydrating the heavier head section. The softened skins are fleshed on a special fleshing machine using a revolving disk knife. This is a highly skilled operation. The skins are then placed in a salt soak prior to tanning. This aids in setting the hair, preventing hair loss.

Depending upon the skins, a degreasing step with solvents and sawdust in a drum may be used. The sawdust is removed using a slatted drum. The tannage is a salt and alum tannage in an oval (race track) vat. After tanning, the skins are centrifuged and dewatered in a drum with sawdust.

If the skins are to be dyed, they are usually dyed in an oval vat with oxidizing dyes. The color is further developed by hanging the skins in air. Fur dyeing formulations are considered proprietary and confidential.

After dyeing and dry drumming, the skins are turned flesh out for greasing. The skins are thoroughly coated with a type of oxidyzed lanolin, then placed in a kicking machine. The kicking machine squeezes and kneads the skin for about an hour. The skins may be drummed with sawdust, caged, turned, and stretched. The process of stretching, fleshing, sawdusting, degreasing, and stretching may follow through several repetitive steps to give maximum

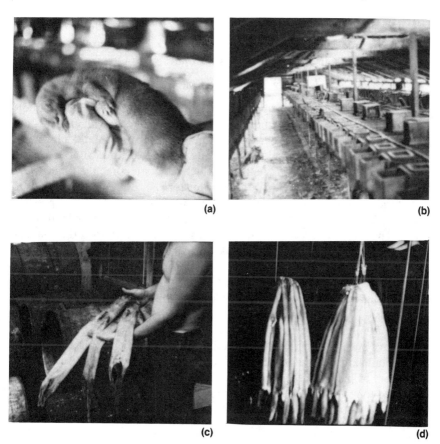

Figure 1   Mink farming. **(a)** The mink are bred to obtain the desired shade of color. **(b)** The individual animals are eventually separated and raised in individual cages. **(c)** The skins are stretched on boards inside out. The flesh can be trimmed away and maximum length obtained. **(d)** Matched skins are grouped together in quantities sufficient for single coats. Mink skins in this form are sold at auction. *Author's photographs.*

length and make the skins as light and soft as possible. Obviously, the more valuable the skin is, the greater the care and effort put into the processing.

In finishing, the skins are stretched, pulled, drummed in sawdust with solvents, caged, and hand cleaned. The skins are then gathered into their individual lots and sent to their owners for sale or processing into garments.

Although mink fur tanning is a specialized skill, the added value of the skin in tanning and finishing is a small percentage of the value of the finished skin. It is unlikely that any new process that would decrease the cost of mink finishing would be of much interest to the established tanners.

KARAKUL

One of the largest factors in the international fur trade is karakul. Karakul or Persian lamb is produced in very dry climates, with the principal sources being Afghanistan and Southwest Africa. Other sources are Russia, Iran, and Pakistan.

Karakul sheep are a rugged breed, well suited to the climate and conditions under which they live. For a major portion of the year in Asia, the grazing areas are in a semidesert condition. The sheep must stand a foraging diet and must have the stamina to go long periods of time without water. The Karakul sheep have heavy fat tails which function much the same as the camel's hump for storage of food and moisture.

A large percentage of the sheep population is owned by nomadic people. These sheep may spend the winter and summer in areas hundreds of miles apart. The ewes are ready to lamb in the spring and will drop their lambs during February, March, and early April. At this time, the sheep are on a starvation diet and in some cases a full term pregnancy will not be accomplished. The stillborn lambs are broadtail and command a premium price. Culling of the flocks will be done at this time in order to obtain broadtails.

When the lamb is born, the fur is in the best possible condition. Where sheep have been on the move and conditions have been difficult, the quality of the lambskin will be better. Male lambs are usually slaughtered within three days of birth for the harvesting of the skin. The females may be kept as needed to replenish the herd. The number of female lambs that are killed will depend upon the anticipated pasture and will be relative to the size of the herd.

Afghan karakul cure system

Salt is spread on the flesh side of the skins in the field, and the skins are piled flesh to flesh and hair to hair with ample quantities of salt. This salt cure method serves to preserve the skins until they are brought to the dealer and sold. The skins are sold individually with a negotiated price for each skin. The skin dealers maintain purchasing centers in the bazaars which are typical storefronts. A large dealer may purchase over 100,000 skins. The skins are removed to a warehouse where they are resalted and kept until the time of the curing season. The cure system involves a washing of the skin in clean water to remove the salt and dirt. The skin is then placed in a vat with a capacity of about 30 gallons. About 80 to 100 skins are placed in a vat (made of camel hide) along with water, salt, and barley flour. The skins are removed regularly, and there is a removal of some of the dirty solution during the curing. Salt and barley flour are added to replenish the vat with a greater portion of barley flour being used. The function of the barley flour in the cure is to provide a source of organic acids through fermentation. The fermentation dominates the process and prevents the growth of harmful bacteria.

Either before curing with the salt and barley flour or after about three days in the cure, the skins are fleshed by breaking off the excess flesh by hand. At the end of two weeks, the skins are removed from the vats, washed gently on the fur side, and air-dried by placing them on a big, hard, flat sand area. The sand will take up heat during the day, and skins placed on it late in the day will dry during the night. Skins are never dried under the direct heat of the sun, and they are moved into the shade during the hottest times of the day.

After the initial drying, the skins are rinsed in water to remove excess salt and dried again on sand beds. The skins are then lightly beaten to remove surface dirt, trimmed, and hand fleshed before sending to a central point for grading.

Southwest Africa cure system

Under the Southwest Africa cure system, the pelts are washed in fresh water immediately after slaughter. They are then trimmed and spread on frames covered with heavy hessian cloth. The skins then air-dry in the shade on the frames and remain in a stretched condition. No salt is used, but a disinfectant or insecticide may be employed in the last wash prior to spread to dry. The dry cured Southwest African skin is smooth and thin in appearance. If the skins are not folded or handled roughly, they will not be damaged in shipping. They are clean to the touch and odorless.

Grading

The cured skins are graded and matched according to color and hair curl into many different grades. A lot of skins is usually sorted twice, and as many as thirty different grades may result. Skins of a particular type and grade are then packed into bales for shipment. The bales are covered with unhaired sheep or goatskin. This protects the skins and prevents bacterial action in the bale. Each bale may contain from 160 to 200 skins, and the number of bales in a lot may be from one to thirty. Broadtail may be packed up to 300 skins.

(a)   (b)

(c)   (d)

Figure 2 Curing Afghan Karakul sheep. (a) Karakul sheep and newborn lambs. (b) Soaking skins in a fermenting acid brine. (c) Drying skins on warm sand beds. (d) Grading skins to uniform quality for auction lots. *Author's photographs.*

Marketing

Karakul skins are sold in auction both in Europe and the United States. Timing is of great importance, since the supply of skins is seasonal. The quality of the skins, the fashion of the time, and economic conditions all play an important role in determining price.

Processing of karakul skins

Soaking procedures vary for the Asian karakul pelts and the Southwest African skins. The African skins are very hard and dry, so special care must be taken to avoid breaking the fiber of the skins. Once soaked, the processing for all karakul is essentially the same. The skins are soaked in a salt solution with mild agitation in a raceway system. After soaking, the fleshing is done on a fleshing machine similar to that used in goatskin or upper leather manufacture. Care must be taken to prevent cutting the flanks. Hand fleshing is often used on the flanks.

After fleshing, the skins are brought back to the wet processing for the chrome tanning. Chrome tanning is used on the processing of karakul skins; alum tannage is only used when particular special effects or bleaching is desired.

Chrome tanning is done in the oval vats similar to those used for soak and the process is somewhat similar to that used in upper leather. It is a very mild process, using relatively dilute liquor and usually takes place over-night. At the end of the tanning process, the fatliquoring oil is added. This is specialty oil which is probably a sulfited oil to make it compatible with the chrome tannage. Simultaneous chrome tannage fatliquor takes place in this procedure. Chrome tanned skins are removed and allowed to drain. The drained skins are then hung on poles to dry. The drying is not done completely, but is done under controlled conditions, so that excessive hard drying does not result.

The dried chrome tanned furs must then have the fur cleaned and degreased. This is done by a process called break drumming. The skins are placed in a drum to which is added about 140 kg of sawdust for 500 skins. With the sawdust is added some water and some solvent, usually a high flash point naptha for the removal of grease. The drum is run approximately three hours at which time the skins are removed. To remove the sawdust clinging to the skins, the skins are tumbled for about a half hour to make the sawdust fall away from the fur.

The moisture content and the solvent that is left in the skins at this point are of great importance, because this controls the temperature and resilience of the skin for the pulling process. The pulling process is a mechanical operation which will soften the skins, open up the fibers, and make soft leather. When the skin is pulled, there should be a visible color change in the fibers and the leather. After pulling, the skins should be quite soft and pliable. Pulling is done on two different types of wheels. One is an overshot wheel and the other is an undershot wheel where the skins may be held against the blade on top and loosened by this action. A large narrow wheel is used more on larger skins such as sheep, goat, and seal. A smaller wheel is used on most other furs, particularly karakul, wolf, fox and others.

The skins are again cleaned by drumming with clean sawdust. An additional pulling and redrumming will follow. In some cases, this process is repeated several times.

The completed skins are returned to the sorters to match the skins into their original classifications as they were received. The skin processor (tanner-dyer) is paid on the basis of the skins returned. He is charged for lost or damaged skins. The processing is a high risk business as are all phases of the fur industry.

SHEARLINGS

Shearlings are sheepskins tanned with the wool intact. Shearlings are the pelts of sheep that have been shorn, and the fleece has not grown out sufficiently to be classed as wool pelts. The skins are grouped in accordance with wool lengths and sold on that basis. It is important in shearlings that the skins be well trimmed. The flesh should not be scored, and the fiber should be in good condition. Shearling pelts may be obtained from almost any sheep pelting area. In the United States, the skins are primarily from the western states. The time of year affects the length of wool on the skin. The skins are salted and cooled with refrigerated air. It is important that the skins not get wet and heated. With the insulating properties of the wool, overheating and bacterial hair loosening can ruin the skin for shearling. The value of the skins is determined by the characteristics of the wool. The quality of the skin is also important for double-faced shearlings (suede on the flesh side).

There are about 20 million shearlings produced per year. This quantity varies with fashion and market conditions. England and Spain each produce about 20% of the world production, Turkey about 15%, while France and the United States each produce about 10%.

Soaking and washing

The processing of shearlings presents some unique problems. The shearlings contain some foreign substances in the wool, including burrs, paint, and natural greases. The removal of these substances must be done with very mild chemical and mechanical action so as not to damage the skins.

The skins are soaked in a paddle system using sodium bicarbonate, detergents, and disinfectants. The skins are washed with several changes of water. A choice must be made between faster cleaning with more paddling or a longer soak.

Pickling and tanning

The skins, after soaking and washing, may be removed from the soak wash water, drained and fleshed, then placed in pickle vats. Paddle equipment is used in this operation to obtain as gentle a mechanical action as possible. The pickling is similar to the usual pre-chrome tanning pickle. The solution is about 5% sodium chloride. The acid (sulfuric) is sufficient to bring the pH of the solution to slightly under 2.0.

The chrome tanning with or without sodium formate may be done in the pickle vat, or the skins may be removed from the pickle vat and drained prior to chrome tanning. The chrome tanning is done in paddle vats for gentle agitation. The chrome tanning may take two days. Some modern processes are done in hide processers with a considerable saving in time and effluent volume.

The development of bisulfited oils has been a great help to the shearling industry. Bisulfited oils may be made stable to the pH and salt conditions found in the chrome tanning bath. As a result, the fatliquoring oils may be added during the chrome tanning. This is the most popular method presently in use.

After chrome tanning (and fatliquoring), the skins are hung to dry. The dry skins are then staked, using a gentle staking action on a machine similar to a setting out machine. Staking was formerly done on an overshot wheel, a machine consisting of a number of dull blades on the periphery of a wheel. The skins were held in the moving blades for staking. The machine, although effective, was time consuming and dangerous, and is in little use today.

The flesh side is then buffed to clean off the flesh and develop a suede surface as desired. The skins are then combed, carded (clipped), and dry cleaned. The combing and clipping machines clean the wool and stand it up so the clippings will result in a uniform wool length. The clipping blade is similar to a reel type lawn mower operating at high speed with great precision. Dry cleaning removes the grease from the wool. There may be a small amount of free oil removed from the skin, but dry cleaning does not destroy the lubricity of the leather. The grease is recovered from the spent solvent and the solvent recycled as in other sheepskin degreasing operations.

Dyeing and finishing

The dyeing of shearlings presents some unique problems. One popular fashion of shearlings is the type in which the wool is kept a natural white and the flesh (leather) is dyed to a desired color as in suede leather. The choice of dyes to accomplish this must be dyes that are fixed by chrome tanned leather, but with no fixation on the wool. Acid dyes with an affinity for the chromium salts (as a mordant) result in a firm fixation of the dye. Good rinsing and removal of the unfixed dye are essential. Fatliquors must also be adjusted.

The final dyeing and finishing of the skins vary depending upon the intended use of the skin. In some cases, the wool is dyed; in others, the leather is dyed and a suede leather made. In some cases, both are dyed.

In any event, the dyeing is done in paddles or modern leather processing machines. The action on the skins is kept very gentle and the liquor ratio high.

After dyeing, the skins are again dried by toggling, conditioned, combed, and brushed to develop a final finish on the leather and wool. The skin may be used for a variety of objects such as automobile seat covers, mitten linings, decorations, rugs, etc. The greatest volume of use, however, is in the manufacture of garmets, suede out and wool in.

SHOE MANUFACTURE

16

The shoe industry is the number one customer of the leather industry. In the United States, which has approximately 6% of the world population, about 20% of the shoes are purchased. Shoe manufacturing consumes approximately 80% of the total leather made. It is essential, therefore, that the leather technologist understand the shoe industry, whether he is working in a tannery, for a chemical supply house, or in research.

ECONOMICS OF THE SHOE INDUSTRY

More than 60% of the footwear made in the United States sells at retail to highly competitive outlets for less than thirty dollars per pair. Both manufacturers and retailers have been very reluctant to disturb fixed price schedules on many lines of shoes. In spite of increasing cost of shoe manufacturing and the decrease in the value of the dollar, shoes remain one of the best consumer buys.

The most significant change in the shoe industry in recent years is the increase in shoe production in many other nations. The development of shoe factories in Korea, Taiwan, and India has been especially important. Shoes are being made for the export market with the close cooperation of the customers in matters of design and quality control. In some cases the factories have been financed by companies in more developed nations by international joint ventures. Most of the major North American and European shoe companies are involved in international manufacture. The manufacturing of shoes is labor intensive and the low cost of labor in the developing nations along with financial incentives such as development bonds and tax incentives can be very attractive.

The shoes that are produced are made in many cases on the most modern machinery and according to the latest fashions. These are definitely not cottage industries. The leather may be domestic or it may be imported, because in many cases domestic hides may not be of the quality required.

There has been a steady increase in shoe imports into the United States which has been a great concern to American manufacturers. It is expected that the trend toward more shoe imports will continue to rise and cause even greater pressure on the domestic shoe industry.

The amount of leather needed for the manufacture of a pair of shoes varies greatly, of course, but the average is slightly over 1 square foot per pair. The cost of the leather, therefore, represents a minor part of the manufacturing cost. The labor, not the leather, is the main determining factor in the production cost of shoes. The shoe industry historically has shown a relatively low profit. About 25–30% of the United States shoe manufacturers routinely report no profits. This non-profit status is attributed to two practices in the industry. For many years the practice of leasing equipment was common. This made it easy to get into shoe manufacturing and out with limited capital investment. The other practice is the selling of non-branded footwear on a rigid, low price line. Fortunately, for the leather and shoe industry, these two practices are becoming less significant. Volume distributors usually sell to chain stores which maintain a low price structure but high sales volume. Job lots of shoes are often sold at reduced prices to retail outlets.

THE MANUFACTURE OF SHOES

The modern shoe manufacturing industry has come a long way in the last century. There has been a transition from the one-man cobbler shop (or ten footers of the eighteenth century) to the highly mechanized production line organization of today. The main differences in the actual production procedures of shoes are in fit and mechanization. In the early cobbler shops, shoes were custom-made by hand; a limited number of shoes were turned out, and styles were few. In modern shoe production there are over 300 sizes and widths of shoes, with many different shapes and styles. Since a manufacturer may have about 50 active styles in his line, and between 80 and 90 sizes of last styles, it is necessary for him to maintain inventory of up to 5,000 different lasts for a style of shoe. (A shoemaker's last is a wooden form over which the shoe is made.) In addition to the problems of lasts, he must maintain efficient production. This can be done with modern shoe machinery, which enables a shoe to travel quickly down a production line.

There are eight production operations in shoe manufacture. The shoes pass in the production line to the different operations (done in different "rooms") in the following order: (1) cutting, (2) fitting or stitching, (3) stock fitting or stock, (4) lasting, (5) bottoming, (6) making, (7) finishing, (8) treeing and packing.

(1) Cutting room. The cutting room uppers are cut with a die by laying the die on the leather and cutting each piece. When cutting the shoes, therefore, it is necessary for the operator to know the characteristics of leather. Definite patterns are laid out by the leather industry to guide in cutting operations. Parts of the shoe that will get severe flexing should be cut from the tightest grain sections of the leather. Parts of the toe and vamp should be cut from the bend areas, and heel parts can be cut from bellies and looser areas. The die should be so placed as to minimize changes in the leather when it is stretched on the last. The cutter must also watch for subtle color differences from one side to another to avoid mismatch. A cutter's job carries a good portion of the profit margin.

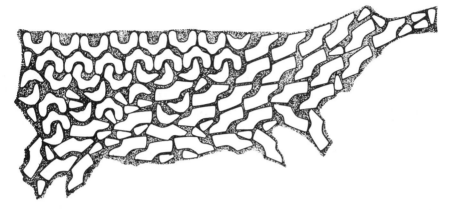

Figure 1   Cutting patterns for side leather uppers. In cutting the shoes from side leather, the parts of the shoe subject to the greatest flexing or strain are cut from the tightest grain parts of the side. The direction of the pieces is also important so that the leather will pull properly over the last.

(2) Fitting or stitching room. The die-cut upper parts of the shoe are fastened together in the fitting room. This is not a simple matter of sewing parts together but is a technique requiring the edges to be skived (or tapered) to get a good bevel for fit. The side upper parts may be further split to a lighter weight for certain parts of the shoe. If there is to be an inner lining, it may be cemented on here. Seams will be rubbed or taped to make a smooth inside. Further reinforcement of the shoe may also be done here. The top edges are folded and cemented to give a smooth edge. Eyelets are added and the entire shoe upper stitched and laced to prepare for the lasting room.

(3) Stock fitting room. In another section of the factory, in the stock fitting (or sole leather) room, the insole or backbone of the shoe is being made. Insoles are die-cut. The outsole blanks are reduced in thickness near the edges.

(4) Lasting room. The lasting operation is a series of forming techniques in which the upper, insole, and reinforcing parts of the shoe are added. Lasting operations vary in different types of construction. An approximate sequence is to attach insoles to the last with small tacks and then to place the upper on the last and tack it; next, the toe is pulled over, using a machine designed for this purpose, and the upper and the insole are either stitched together, stapled, cemented, as the construction demands.

(5 and 6) Bottoming and making room. The bottom and making room are taken together since one relies on the other, and the two procedures are frequently done together. These processes sequentially attach the sole, attach heels, trim, and finish the shoe. The filler materials, if used, also are added

here. The differences in individual manufacturer's choices and methods are most evident in these two rooms.

(7) Finishing room. With the sole attached, the outsoles are trimmed and burnished. Edges are stained and waxes applied and buffed. After the bottom finishing, the shoe is removed from the last. In men's shoes the heel is usually attached before removing from the last, but in women's shoes with high heels, the nails are driven from the inside down into the heel to hold it in place.

(8) Treeing and packing room. The finished shoes are cleaned and in some cases given a top dressing (treeing). The shoes are buffed, inspected, and packed for shipment.

SHOE DESIGN

There have been great changes in the methods of shoe design. The interest in walking and running as exercise and competitive sports has sparked elaborate studies of shoe comfort and foot support. The development of computer-aided design (CAD) techniques has been used by the shoe industry extensively. Sport shoes and dress shoes are now more comfortable than ever before.

SHOE CONSTRUCTION METHODS

Shoe construction methods vary chiefly in the method of the attachment of the sole. These methods are quite different and are the main determining factor in the cost, durability, style, and use of the shoe.

McKay and Littleway system

In the McKay system the upper is fastened to the insole by means of tacks which have the heads turned over by a steel plate held in the bottom of the last. The ridges of the shoe are ruffed, the filler is added, and the leather sole chain stitched on. The stitching is hidden from the bottom of the sole by skiving the sole. Another system called the Littleway method, is also done on a flat-lasted shoe. The upper is fastened to the insole without skiving and the chain stitch is replaced by a lock stitch of more durable construction.

Sliplasted shoe

A sliplasted shoe has the advantage of inexpensive production. It can be used only where the upper and sock linings are accurately cut. In the sliplasted shoe the upper and the sock lining (light insole) are stitched together prior to

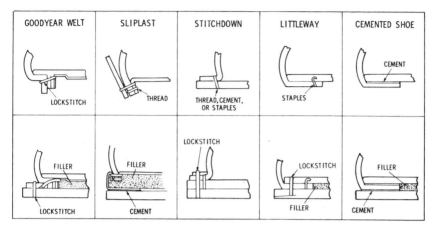

Figure 2 Five common methods of attaching the sole are shown here. The importance of insole materials and upper leathers in their ability to hold the stitch can be
seen from these drawings.

being placed on the last. Once the sock lining is in place on the last, filler
and inner layers are added on the outside and fastened by cement. Open-
toes and platform shoes are more easily made by sliplasted procedures.

Stitchdown shoe

The stitchdown shoe differs from the Littleway in that the insole is fastened
to the upper by flairing the upper outward rather than inward. In this condition the stitching is close to the last with margins on the upper and the lining.
The method is popular in children's shoes and is one of the easiest methods
of manufacture.

Moccasin

The moccasin and the loafer are very popular types of shoes. The insole and
the upper are made of one piece of leather which wraps all around and is
stitched at the top of the vamp. For this upper, heavily vegetable re-tanned
leathers are used. The shoes are soaked in water to aid in the forming of
compound curves. They are then stitched by hand on the last and the cut
sole is lockstitched to the bottom.

Goodyear welted shoes

This shoe is recognized as a high quality shoe of the construction used in
heavy men's shoes. The insole is cut and skived, and the upper is stitched to

the tab formed by the two skived edges on the insole. The welt is stitched into this joint, and after fillers are added, the cut sole is lockstitched.

Cemented shoe

For high-speed production and low cost, many shoes are made by cementing. The shoe is assembled with adhesives, formed over a last, and the upper is fastened to the insole by means of partly penetrated staples or tacks. The bottom is skived and the sole is tapered for a tight fit. Some of the adhesives are activated by heat, others have 24 hour tack. Cement is applied by high pressure and in a clamp to assure good adhesion of the sole.

Direct-molded sole

A recent development based on the new synthetic rubbers is the direct-molded sole. The upper and insole of the shoe are formed, and the sole is cut to the desired shape. A mold is placed around the sole and thermostatically controlled heat softens the sole and compound. Pressure applied then will force the sole into the desired shape. Chemical changes taking place in the sole result in vulcanization of the sole with the upper material for a firm bond. This method is most popular with sport shoes and for use on heavy leather boots.

Injection-molded sole

A variation of the direct-molded sole is a process in which the shoe upper and insole assembly are placed in a mold with the desired shape of the outsole. The outsole material is then formed by injection molding into the die. Both direct-molded sole with vulcanized shoe and molded shoe are adaptable to low-cost, high-speed methods.

LOOKING AHEAD

The shoe industry, like other soft goods industries, has not enjoyed an increase in consumption with better times. The per capita purchase of shoes in the United States has remained constant in times of economic stress as in times of plenty. Coupled with this, the American shoe industry is plagued with the problem of imports. In order to maintain their position in the competitive market with each other and from abroad, domestic shoe manufacturers can be expected to turn more toward lower cost production and higher quality. The shoe manufacturer will expect more from leather in terms of compatibility with his processes, as well as more in performance and durability. The expansion of the direct-molded sole and injection molded sole shoe is already placing requirements on the leather with regard to oil content and adhesion. It behooves the tanner to produce leathers of good quality, new processes, compatibility, mechanical durability, and above all, unique leather qualities that are plainly evident to the public.

The shift of the shoe industry production to overseas has resulted in the closing of many shoe factories in the United States. The shoe companies must meet the challenge of low costs by better quality. New designs and the skillful use of combinations of materials are essential to avoid being caught in a price bind. The tanner must also create new leather fashions and improve the performance of the leather to succeed in the changing competitive market.

REFERENCES

"How American Shoes are Made," Boston, United Shoe Machinery Corp., 1966.

Maeser, M., *J. Am. Leather Chemists' Assoc.,* **58**, 456 (1963).

Moynahan, F.G., *Leather and Shoes* (February 4, 1967).

Prelcec, D., *Kozaiobuca,* **7**, 185 (1967).

Riley, J.J., "U.S. Industrial Outlook," p. 75, Washington, D.C., 1966.

U.S. Department of Commerce, USCOMM-DC-(1963).

U.S. Department of Commerce, USCOMM-DC-(1966).

LEATHER SUBSTITUTE
MATERIAL (MAN MADES)
17

This chapter will discuss leather substitutes as a point of interest to the leather field. A complete set of descriptions is not possible because imitation leathers are changing so rapidly.

Probably no subject is discussed more thoroughly by persons in the leather industry than the relative merits of man-made materials and real leather. Prior to Word War II no materials were available that were serious competitors to the leather industry. Today, however, man mades are available with properties similar to those of leather in practically every category of leather production. The tanner's opinion of the relative success of the various synthetics, on a short- and long-range viewpoint, is vital to his decisions regarding investment and the direction of his technical developments.

In gathering information on the relative values of leather and leather substitute materials, we can see the problem from several points of view. We can physically test and evaluate leather, we can run both physical and chemical tests, and we can clinically test the properties for evaluation. We can rely also on the opinions of shoe manufacturers with regard to performance and what they expect of the leather industry.

The evaluations and test methods employed by leather technologists in proving the value of leather can be disputed by the manufacturers of man mades. The personal prejudices of the people making the tests must be weighed when considering the validity of those tests. For example, there is a lack of communication between the shoe manufacturer and the leather manufacturer with regard to quality and performance which only accentuates the invalidity of these tests. Leather and man mades should be judged on their qualities as needed by the particular manufacturer. For the manufacture of shoes, distinct mechanical and physical properties are required of the materials in order to assure proper fit and shoe comfort. Mechanical strength and flexibility are needed to survive the shoe manufacturing process. Once the shoe is made it must adjust to the wearer's foot; it must be neither too warm nor too cold; it must resist the effects of water, abrasion, and temperature, and still be an item of beauty.

MAN-MADE MATERIALS

The characteristics of leather have proven to be very difficult to reproduce. The popularity of the man-made materials in shoes has been due greatly to the easy care characteristics of the man mades.

The man mades are made on continuous roll machines. The use of transfer coating, roll coating, and continuous printing can be accomplished in an efficient and uniform manner. The substrate may be coated (several coats), laminated and cured in a single pass. Since the substrate can usually stand greater temperatures than leather, resin cure systems may be used that would not be applicable on leather. These manufacturing techniques also eliminate much of the waste of materials encountered in leather finishing due to the irregular shapes of the leather. The man mades therefore, have advantages in manufacture that give some surface characteristics that could not be obtained on leather. This can be done at a lower cost than on leather. We can expect that the man mades will continue to be produced at a price lower than leather if but for these reasons alone.

The cost advantages of the man mades have shown, as expected, to be most significant in the lower priced leathers and in applications involving large surface areas. This is evident in garment and upholstery leather fields. Acceptance of the man mades in these areas has been widespread.

Leather substitute materials or man mades have made great strides in the past few years. These have resulted in materials of appearance closer to leather in some cases. In other cases, the man mades have stopped trying to look like leather and have set their sales appeal on automation of production. Complete shoes can be made by injection molding in a way to replicate exactly the leather shoe it copies. Garment materials of foamed urethane on fabric are well accepted.

With these advances, however, there has not been a decrease in leather demand. The man mades have in some cases emphasized the valuable properties of leather and thus have helped leather's marketing position. Golf shoes a few years ago were predominantly manufactured from man-made materials. Now, leather again has the quality market. There is a great demand for full grain garment leathers. There is no substitute for this material in the eyes of the public. Several multi-million dollar investments have been made in the development of high quality man-made upper materials. The results of these efforts have in most cases resulted in technical and financial failure.

The development of the world wide energy shortage and its accompanying increase in oil prices, has also been a factor. With higher relative oil prices, petrochemicals have also increased in price. The net result was an increase in the price of the man mades relative to the price of leather.

The statements made in the "Looking Ahead" section of this chapter of previous editions of this book have proven to be true during the past 6 years. The author believes that the statements made then are true now and will continue to be current for a reasonable time in the future.

Leather should hold its position on performance and quality rather than attempt to compete with synthetics on a cost basis. In a John Arthur Wilson Memorial lecture before the American Leather Chemists' Association, M.

Maeser outlined the requirements of leather as a shoemaking material. The procedure of making several types of shoes was given in detail, and structural problems were pointed out. Maeser also included a study of the behavior of leather in adjusting to the wearer's foot.

In shoemaking, the skiving and beveling of the edges and the ability of leather to be tapered and hold a stitch without raveling is paramount. Manmade materials do not have these qualities as yet.

In the lasting of a shoe, the leather is pulled over the toe in a compound curve. The three-dimensional fiber structure of leather and its ability to stretch and compress upon itself to form a compound curve is important in shoemaking. Leather is easier to last as it will shrink to the last when heated to remove wrinkles. Once the shoe is in shape and has been worn, it must hold this shape but with enough natural distortion to adjust to the act of walking and to the individual foot. Synthetic materials have been processed to return to shape. Referring again to Maeser's lecture, leather characteristics were further outlined: "Leather upper has many properties and characteristics that can be used to advantage in shoemaking, and these properties are important to wear. The capacity of leather to be either stretched or compressed into its own voids without altering its surface appearance, its ability to be skived without losing its strength, its ability to be stitched close to unturned or unbound hitches, the fact that its edges do not fray and tear in wear, its lack of notch sensitivity, its great fatigue resistance, its multiple layer fibrous structure and rich, fine porous grain, its beauty and texture which a person can enjoy stroking with his hand and rubbing against his face, its ability to be wet and dry with no loss of character, its moldability and capacity to shrink and set permanently, and its lasting shape are all good shoemaking properties. This is an impressive list of fine properties, and no flexible sheeting material ever made has so many desirable characteristics built into it."

The advantages cited for the man-made materials are as follows: (1) high cutting yield cited at 10–40% savings; (2) multiple cutting with cutting dyes as high as eight deep; (3) less skilled cutters needed although the material must be cut directionally; (4) more uniform stitching; (5) less damage during manufacture; (6) faster processing (no mulling is needed); (7) easier processing; (8) more uniform quality shoes; (9) opportunity for automation.

It is interesting to note that the advantages of the man-made upper materials are all oriented toward ease of shoe manufacture. The development of segmenting hides into a more uniform size and shape will possibly strengthen the position of leather in this respect.

Comfort factors of leather are hard to define. Laboratory tests on air permeability, moisture permeability, moisture take-up, etc. are difficult to assess. Detailed studies have been made by the Canadian Military Laboratories to compare military boots made with leather to those made with synthetic soles, uppers, and linings. Thermocouple measurements were made of the temperature of the foot at various points and clinical opinions were obtained from the wearers. The interesting part of this study is that the temperature of the foot was lowered due to the wicking action of the perspiration of the foot through the leather in evaporating on the surface. The take-up of moisture and permeability to water vapor are not the same thing. Shoe comfort is related more to the take-up of liquid moisture and evaporation than to the passage of air or water vapor. Porosity and good air permeability do not necessarily mean comfort and good cooling of the foot. This factor, along with the ability to shape to the foot, has been the main reason for the choice of leather for shoes.

The characteristics of the leather substitute materials used for shoe uppers can be appreciated by observing a number of physical tests. By means of these tests we can estimate quite well the value of a particular material as a shoemaking component. A typical set of data comparing some synthetic leather upper materials and leather is given in Table 1.

Tensile strength. Tensile strength can be expressed in terms of "force per unit cross sectional area" as commonly done in engineering materials, or it can be expressed in "force per inch of width" as it is in textiles. In the shoe upper materials all of the materials are approximately the same weight (thickness) and the data is directly comparable. The synthetic upper materials are directional, having different characteristics of ultimate tensile strength in the length and width direction. Leather also has a difference in these characteristics, depending upon the section of the hide from which it is cut. These synthetic materials do not approach leather in tensile strength.

Elongation at the break. The ultimate stretch of the upper material at the time of the break is easily measured on any of the standard test machines. Here again, the directionality of the synthetic materials is evident. There is a great difference in the flexibility in the synthetic materials. Some of them may be stretched as little as 20% of length at the time of the break, and others may be stretched as much as 150%. Synthetic materials are usually reinforced with cloth in addition to the non-woven mat. If the strength of the material is due primarily to the cloth reinforcement, stretch will be greatly decreased. A synthetic upper material without cloth reinforcement will generally have greater elasticity.

Modulus. Modulus, as applied here, is the force per unit inch width necessary to increase the length of the sample by 10%. Here we see that leather may have a relatively low modulus in spite of its large tensile strength and relatively small elongation at the break. Synthetic materials vary widely depending on the material being tested and its method of construction. The modulus is very important to foot comfort. Without a perfect fit, a material that has a high modulus will cause constant pressure on the foot. Leather, particularly as it is broken in, will shape to the foot, and the initial stretch will take place with a permanent set. This is a main factor contributing to the comfort of leather.

TABLE 1
Physical Characteristics of Leather and Synthetic Materials.

Material	Tensile strength (lb/in.)		Elongation at break%		Modulus (lb/in.)		Stitch tear (lb)		Moisture-vapor permeability (g/M²/hr.)	Moisture absorption at 100% humidity
	Length	Width	Length	Width	Length	Width	Length	Width		
A	82	77	34	40	51	32	22	30	22.5	<1%
B	55	81	110	68	6	15	26	33	17.1	<1%
C	86	100	18	21	52	50	30	25	4.2	3%
D	100	103	12	32	75	32	34	33	17.4	3%
Upper leather	185		60		25		70		17.6	15%

Stitch tear test. The ability of a material to hold a stitch is of great importance. Good leather shows an excellent ability to hold stitches. The man-made upper materials vary considerably in their ability to hold stitches.

Moisture vapor permeability. It is important to the comfort of a shoe that the moisture be taken from the foot, carried through to the surface of the leather, and evaporated. Moisture vapor permeability, therefore, has been considered to be of importance in both man-made and leather upper materials. The value of the moisture vapor permeability will depend primarily on the finish. Those materials having a sealed finish coat or film of an impervious resin will have very limited moisture vapor permeability.

Moisture absorption. If we equilibrate the synthetic materials and leather with moisture and various relative humidities, we can see that there is a great difference between leather and synthetic upper materials in their ability to take up water. It is not enough to have moisture vapor permeability, moisture from the foot must also be carried by the fibers to the surface of the leather to be evaporated. The man-made upper materials show very little tendency to absorb the moisture from a humid atmosphere. This is an important comfort factor.

LOOKING AHEAD

Statistically, the leather industry in the United States faces strong competition from several sources: the synthetic uppers made of straw and textiles, imitation leathers, and imports. These sources will make inroads into the leather markets, whether they be for shoes, handbags, luggage, upholstery, sporting goods, garments, etc. We have confined our discussion to the shoe industry since it is the most critical and the most current. The difference between the total world supply of leather-making materials and the demand for leather-like products must be made up by the production of synthetics. The characteristics of leather are unique, and in spite of the improvements being made in substitute materials, it is unlikely that it will be more profitable to turn leather-making raw materials into food, fertilizer, or industrial use. Plastics have produced imitation billfolds, but reptile leathers remain expensive and in demand. Plastics and textiles have also done a profitable business in the garment field without driving the grain and suede tanners out of business.

The future of leather, in all categories, depends upon the leather technicians who will produce a leather of beauty and quality that stresses the unique characteristics of leather. The newest standards of leather performance with regard to uniformity of cutting, water repellency, dimensional stability, comfort and wear, and consistent reliability in performance must be developed. To those forward-looking leather producers will go the rewards of continued profitable business.

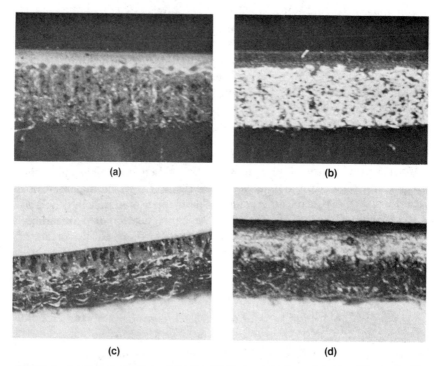

(a)   (b)

(c)   (d)

Figure 1   Leather substitute material. (a) Non-woven base fabric reinforced with a finish and a top dressing; this is a material for shoe uppers. (b) Non-woven base with a finish; shoe uppers. (c) Non-woven base with a foam finish for shoe linings. (d) Side upper leather. *Author's photographs.*

There will be an increase in leather production in the developing nations; and there will continue to be a brisk demand for leather and leather-making materials. The industry will operate under entirely new standards of economics, technological acceptability, sales policies, and raw materials purchasing. The management people who can anticipate these changes, and innovate rather than follow, will advance a great industry.

REFERENCES

Boswell, J. Mc., *J. Soc. Leather Trades' Chemists*, **48**, 488 (1964).
Maeser, M., *J. Am. Leather Chemists' Assoc.*, **58**, 456 (1963).
Roddy, W. T., *Boot and Shoe Recorder* (September 1, 1966).

TANNERY EFFLUENT

18

Tanning has the unenviable reputation for being one of the filthiest, evil smelling of industries. This reputation was established by the character of the village tannery a century ago, and still persists. With the expansion of the industry to modern, highly mechanized factories in urban areas, the problem became more acute and the reputation of the industry deteriorated further. Today, with the population explosion and the strain being put on our world for saving our natural resources it is becoming more apparent that the disposal of wastes is a matter of the industry's responsibility to the society around it. The pollution control regulations which were relatively loose a few decades ago are now extremely tight. Efforts are being made to limit and eliminate as much as possible the discharge of polluting materials into the atmosphere, ground and primarily, the discharge into the water. The leather industry is very much in the balance and the efforts of a given tannery now and in the future will be centered more on solving the pollution control problem than any other single factor. The costs involved are tremendous and could wipe out the profits of a given tannery for a time. The regulations of pollution controls have resulted in the closing of a number of tanneries in the United States.

EFFLUENT CONTROL REGULATIONS

The Environmental Protection Agency (EPA) has studied various high polluting industries and made surveys, through consultants, to determine what the best practical treatment of industrial wastes can be. On this basis permits are granted to industries to discharge wastes in accordance with their production and what is required to get satisfactory discharge levels by the best practical treatment. The levels of these requirements are still under discussion but the concepts are well established.

Direct discharges, those discharging directly into streams, must treat the wastes for the removal of toxic materials and decrease the suspended solids and biochemical oxygen demand (BOD) to an acceptable level.

Indirect discharges are those discharging into municipal systems. They may be required to decrease the toxic materials to an acceptable level. Federal regulations require specific levels of treatment from a tannery including sulfide and chromium. Some of these require treatments which can be done easily, but some must be done with great difficulty. State and local requirements may be more strict.

TABLE 1

Estimated Pollution Load from a 1,000 Hide/Day Cattlehide Tannery* 60 lb Green Salted Cattlehide

		lb/1,000	*lb/day*
Soak—50,000 gal/day			
BOD		15	900
Total Suspended Solids		25	1,500
Total Solids		200	12,900
pH	6.0–8.0		
Unhairing—50,000 gal/day			
BOD		50	3,000
Total Suspended Solids		65	3,900
Total Solids		150	9,000
Oil & Grease		10	600
Sulfide		7.5	450
pH	11.0–12.5		
Lime—100,000 gal/day			
BOD		10	600
Total Suspended Solids		15	900
Total Solids		50	3,000
Oil & Grease		5	300
Sulfide		1	60
pH	11.0–12.5		
Bating—60,000 gal/day			
BOD		6	50
Total Suspended Solids		7	420
Total Solids		5	1,200
Oil & Grease		5	1,200
Sulfide		—	—
pH	7.0–10.0		
Chrome Tanning—15,000 gal/day			
BOD		4	240
Total Suspended Solids		5	300
Total Solids		150	9,000
Oil & Grease		10	600
Chromium		5	300
pH	3.5–4.0		
Retan, Coloring & Fatliquoring—100,000 gal/day			
BOD		2	120
Total Suspended Solids		3	180
Total Solids		20	1,200
Oil & Grease		10	100
Chromium		1	60
pH	4.0–5.0		

*Average data based on Thorstensen Laboratory studies of 12 tanneries.

294

TABLE 1 (cont.)

	lb/1,000	lb/day
Finishing—50,000 gal/day		
BOD	3	180
Total Suspended Solids	5	300
Total Solids	8	480
Oil & Grease	10	600
pH	5.0–8.0	

Soaking

A major portion of the tannery wastes comes from the early part of the production process. In the manufacture of chrome tanned leather from cattlehide, the initial stages of soaking result in a heavily polluted water containing salt from the curing, soluble proteins and manure. The water may vary considerably from one tannery to another, thus the concentration of the effluent operation varies. In any case, the effluent is dark brown and very dirty, but has very little objectionable odor which indicates that little or no bacterial action has taken place on the hides. The soaking operation will result in wastes as listed in Table 1.

The total quantity of wastes that are discharged from the soaking operation is entirely from the hides and not the method of the soaking operation. The total solids discharged will vary up to 250 lbs of solids per 1,000 lbs of hides, most of which is salt.

Unhairing

Concentrated unhairing wastes are the largest problem of the effluent of a tannery. The BOD produced represents approximately 50% of the total BOD produced by the tannery and the volumes of water are quite large due to the washing that is necessary. The total BOD produced is higher in the pulping process since there are more soluble proteins present.

Sulfide

Sulfide is a most difficult material in that it is highly toxic and has an obnoxious odor. When sulfide is released into municipal sewer systems, the accumulation of sulfide gases in the pipes causes corrosion and may also release poisonous gas in the low parts of the sewer. Lime is also objectionable, because it clogs sewers and leaves an excess of alkalinity in municipal sewer plants.

Bate, pickle and tan operations

Ammonium salts are used to delime the complex to remove the lime from leather. This will contribute relatively large quantities of ammonium salts and some soluble proteins, suspended solids as well as some lime to the effluent. The deliming wastes will be fairly close to neutral pH. BOD will be relatively low. Part of the objection will be the oil and grease and the ammonium salts that are present.

Pickle liquors contain large quantities of salt in the amount of 50,000 mg per liter and contribute strong acids. In modern tannery production there is no dumping of the pickle liquors so it is not a factor in the effluent problem. The small amount of pickle that may be lost to the sewer is usually insignificant relative to the lime from the unhairing wastes.

Chrome tanning

Chrome tanning wastes from the tanning operation will contribute approximately 5 lbs of chromium per 1,000 lbs of hides (assuming a normal fixation of the chromium in the tannage). Trivalent chromium is used in tanning and is not considered toxic as it is apparently not an accumulative poison. Hexavalent chromium (not present in tannery wastes) on the other hand is very toxic. Sewage disposal plants have been operated with chromium from tanneries present in the wastes with no objectionable results. However, the EPA requirements on the discharge of industrial wastes require the essential total removal of chromium by precipitation.

Coloring and fatliquoring

Coloring and fatliquoring wastes consist of dyes, vegetable tanning materials, syntans and other specialty chemicals, some fatliquoring oils and some of the natural oils from the hides and skins themselves. These wastes, although highly colored, contribute relatively small quantities of objectionable BOD, suspended solids and contribute relatively small quantities of chromium wastes.

Finishing

In the finishing operation, the wastes may contain some suspended resins and pigments. It may also contain some of the solvents accumulated in the wash down system.

The total wastes given on a side leather tannery were given in a survey by the New England Water Pollution Control Commission and are ás listed in Table 2.

TABLE 2

Average Composition and Contribution of Strong Liquors.*

	Volume		BOD				Sodium chloride ppm	Total hardness ppm	Protein ppm	Total solids ppm	Volatile solids ppm
Process	Gallons per day	% of total	ppm	Pounds per day	% of total	Pounds per 1000 lb hide					
Soaks	73100	42	2200	1310	20	15[a]	20000	670	1900	30000	3600
Unhair	27200	16	15500	3510	52	40[a]	18000	25000	22700	78000	18000
Relime	27200	16	650	147	2	2[a]	3500	25000	—	20300	2500
Delime and bate	17600	10	6000	880	13	10[b]	<10	4100	4300	15000	8800
Pickle	9800	6	2900	237	4	3[b]	47000	2400	—	79000	7200
Chrome tan	8500	5	6500[d]	425	6	8[b]	26000	1800	—	93000	13000
Color and fatliquor											
First dump	5100	3	2000	85	1	3[c]	—	—	—	16000	8000
Second dump	5100	3	2200	93	1	3[c]	250	2600	—	9500	4900
Total	173600	—	—	6687	—	—	—	—	—	—	—

[a] Based on wet, salted hide.
[b] Based on fleshed, split hide, after re-lime.
[c] Based on chrome tanned leather.
[d] Estimated at 50% of volatile solids concentration.
*Based on a study by the New England Interstate Water Pollution Control Commission; the data are typical for a tannery working approximately 80,000 lb of cattle hide per day.

Sole leather tanning

The wastes from sole leather tanneries are similar to those from a chrome tanned side leather tannery with regard to the discharge from the soak. The liming process used in the sole leather tannery results in the hair save system and therefore the discharge contains less fibrous material, lower BOD, and somewhat more dissolved protein. The vegetable tanning process results in quantities of liquors that can be recycled but they do contribute to high suspended solids and high BOD, so their ultimate disposal is a problem.

Light leather

In the light leather field, such as the production from pickled sheepskins, the pickled sheep are degreased and a solvent recovery system must be used. Other than that, the chrome tanning and dyeing, fatliquoring operations contribute wastes similar to those found in the side and upper leather tanneries.

Split leather

In the manufacture of split leather, the same problems occur as occur in the coloring and fatliquoring operation because the same materials are used and in approximately the same type of formulations.

Pigskins

In recent years, there has been an increased interest in pigskins as a raw material. The manufacture of pigskins is limited so only a small amount of data is available on the processing and the effluent that results. We would anticipate that the operation of a pigskin tannery would be similar to that of a sheepskin tannery where solvent degreasing systems are used.

EXISTING RAW WASTE LOADS

The Environmental Protection Agency has studied extensively the raw waste load resulting in various classifications of tanneries. The data in Table 3, published in 1979, are given for each of the six classifications of tanneries studied. We have included here only the data given in kg per 1000 kg of hides processed. An inspection of the data is most helpful for comparison of the sources and anticipated polution load in any particular tannery.

The difference between category 1 and 2 is the type of unhairing system. Category 1 is hair pulp and category 2 is hair save. Hair pulp results in higher BOD, chemical oxygen demand (COD), sulfides and total nitrogen. Suspended solids and oil and grease are not significantly different by the two systems.

The difference between category 1 and 6 (through the blue) is slight, indicating most of the effluent comes from unhairing and tanning wastes. This is also confirmed by comparison with category 4 (retan wet finish); category 6 plus category 4 approaches category 1 in total load.

Category 3 (non-chrome tan hair save) shows somewhat higher (COD) than category 2 due to dissolved vegetable tanning material. All other categories are about the same or less, due to more mild processes and some recycling. The high sulfides in category 3 are the exception.

The present U.S. policy

Under Federal statute every industry discharging directly into a stream must receive a permit from the Federal Government and the state which prescribes a discharge that is allowed. It also prescribes a schedule of compliance for the discharge in accordance with the standards set by the Federal Government. The standards required on an individual tannery may not be the same in one area to another due to the influence of the state. The industry guidelines of the various categories are listed in Table IV.

It is quite evident that tannery wastes in all cases must be essentially free of sulfide, free of chromium and the discharge of suspended solids, BOD, oil and grease must be kept to a minimum and the pH must be adjusted to the acceptable range. The total flow of water is not of particular concern since the quantity of material discharged is limited in terms of lbs per day. For indirect discharges, treatment of the toxic wastes are required either by the local community, state or the Federal Government. Cost recovery charges will be made by the local governing body, based on flow, suspended solids and BOD, by which a tannery must pay its full fair share of construction and also operation and maintenance costs.

PROCESSING OF TANNERY EFFLUENT

Several detailed studies have been made of individual tanneries concerning the quantity of effluent material discharged during manufacture. On reviewing the data obtained in some of these studies, one is struck with the fact that the quantity of water used varies greatly from one tannery to another. The total quantity of chemicals used for a given quantity of hide, however, will vary to a much lesser extent, depending upon the type of leather being produced.

TABLE 3

Level Of Treatment—Waste Stream Composition

Category 1: Hair Pulp—Chrome Tan Retan-Wet Finish

Pollutant or Pollutant Property	Existing Average Raw Waste Load (Parameter mg per l/ lb per 1000 lb)
Flow	4.6 (gal/lb)
Biochemical Oxygen Demand BOD$_5$	1620/62.3
Total Suspended Solids (TSS)	2410/92.3
Chemical Oxygen Demand (COD)	4640/178
Oil and Grease	400/15.4
Total Chromium	75/2.9
Sulfide	64/2.47
Total Kjeldhal Nitrogen (TKN)	330/12.6
Ammonia	104/3.98
Phenol	1.0/.038

Category 2: Hair Save—Chrome Tan Retan-Wet Finish

Pollutant or Pollutant Property	Existing Average Raw Waste Load (Parameter mg per l/ lb per 1000 lb)
Flow	5.5 (gal/lb)
Biochemical Oxygen Demand BOD$_5$	908/45.1
Total Suspended Solids (TSS)	1920/88.3
Chemical Oxygen Demand (COD)	2610/119.7
Oil and Grease	244/11.2
Total Chromium	31/1.4
Sulfide	20/0.92
Total Kjeldhal Nitrogen (TKN)	137/6.3
Ammonia	90/4.15
Phenol	2.2/.1

Category 3: Hair Save—Non Chrome Tan—Retan—Wet Finish

Pollutant or Pollutant Property	Existing Average Raw Waste Load (Parameter mg per l/ lb per 1000 lb)

Category 4: Retan—Wet Finish

Pollutant or Pollutant Property	Existing Average Raw Waste Load (Parameter mg per l/ lb per 1000 lb)

TABLE 3 (cont.)

Pollutant or Pollutant Property	Existing Average Raw Waste Load (Parameter mg per l/ lb per 1000 lb)
Flow	4.0 (gal/lb)
Biochemical Oxygen Demand (BOD$_5$)	1180/39.2
Total Suspended Solids (TSS)	1680/56.1
Chemical Oxygen Demand (COD)	5120/170.9
Oil & Grease	340/11.3
Total Chromium	11/.38
Sulfide	68/2.3
Total Kjeldhal Nitrogen (TKN)	202/6.8
Ammonia	90/3
Phenol	1.2/.04

Category 5: No Beamhouse

Pollutant or Pollutant Property	Existing Average Raw Waste Load (Parameter mg per l/ lb per 1000 lb)
Flow	3.3 (gal/lb)
Biochemical Oxygen Demand (BOD$_5$)	1000/27.6
Total Suspended Solids (TSS)	630/17.4
Chemical Oxygen Demand (COD)	1700/47
Oil and Grease	340/9.5
Total Chromium	68/1.9
Sulfide	3.2/.09
Total Kjeldhal Nitrogen (TKN)	170/4.6
Ammonia	36/1
Phenol	1.2/.034

Pollutant or Pollutant Property	Existing Average Raw Waste Load (Parameter mg per l/ lb per 1000 lb)
Flow	1.7 (gal/lb)
Biochemical Oxygen Demand (BOD$_5$)	780/11
Total Suspended Solids (TSS)	820/11.6
Chemical Oxygen Demand (COD)	3120/44.2
Oil & Grease	270/3.9
Total Chromium	53/.75
Sulfide	1.1/.015
Total Kjeldhal Nitrogen (TKN)	210/3
Ammonia	110/1.55
Phenol	3.9/.055

Category 6: Thru-The-Blue

Pollutant or Pollutant Property	Existing Average Raw Waste Load (Parameter mg per l/ lb per 1000 lb)
Flow	2.70 (gal/lb)
Biochemical Oxygen Demand (BOD$_5$)	2280/51.3
Total Suspended Solids (TSS)	3580/80.7
Chemical Oxygen Demand (COD)	5900/133
Oil and Grease	520/11.6
Total Chromium	98/2.2
Sulfide	110/2.5
Total Kjeldhal Nitrogen (TKN)	430/9.6
Ammonia	110/2.5
Phenol	1.3./.03

Table taken from U.S. Environmental Protection Agency, *Proposed Effluent Limitations Guidelines, New Source Performance Standards*, Washington, D.C., July 1979.

There are two methods of approaching the effluent problem: Either the process must be changed so that the objectionable chemicals can be omitted, or methods of treating the effluent must be devised. Major changes in processing are not easily accomplished unless the process results in an increase in the value of the leather with a negligible increase in cost. Considering the cost of leather and the materials going into it, it would be unrealistic to expect major processing changes to be made strictly on the basis of decreasing the effluent problem.

In considering possible methods of processing tannery effluent, the first information necessary is the quantity and type of effluent being created by a particular tannery.

The second necessary consideration is the location of the tannery and its relationship to the community. A small tannery located in a large city served by a municipal sewage treatment plant may have no difficulty at all; the discharge of its chemicals may be lost in the general preponderance of community effluent. The isolated large tannery operating near a small stream, however, is faced with an entirely different set of problems. A further complicating factor is that the discharge from the tannery is not continuous throughout the day. The dumping of tanning drums and lime paddles occurs in batches, and the type of effluent may be very strongly polluting at one minute and quite innocent a few minutes later.

The nature of tannery effluent varies greatly from one part of the process to another. Only a small percentage of the total volume of the effluent may contribute the greatest amount of objectionable material.

Flow

Considering the treatment of tannery wastes, the first thought that comes to mind is the question of flow of quantities of water used in tanneries in all phases of the wet operation. The water used through the formulations in the tannery is seldom over 60% of the total water consumed by the process. There are numerous leaks and drippings throughout the plant which contribute greatly to flow. Conservation efforts of a tannery begin with fixing leaky valves, etc. An area of improving the flow is by decreasing the washes and controlling proper pressure in the pipes. Ratio and flow volumes must be carefully controlled.

Suspended Solids and BOD

Both the suspended solids and BOD in tannery effluent are primarily from the unhairing. The total quantity of material discharged is independent of the process used. This is more dependent upon the hides themselves. Various levels of discharges, suspended solids and BOD are not decreased by decreasing the total volume of flow. Of the total solids discharged from the tannery, approximately 50% are due to salt, either from the salt cure of the hides or from the chrome tanning. The removal of salt from the tannery wastes has not been considered to be feasible in the United States. Little attention has been given to this. It is anticipated that some effort will be made for the removal of salt sometime in the future.

Tannery waste treatments

The treatment of effluent from a tannery is generally designed to be operated in three distinct steps: screening, coagulation, and settling.

a) Screening. The removal of hair and small suspended hide particles can be accomplished by a mechanical screening system. These are usually rotating, self-cleaning screens which will take out particles larger than one-sixteenth inch in diameter.

b) Coagulation. The materials of a tannery will interact mutually. For example, the acid of the pickle will be neutralized by the alkalinity of the lime. The alkalinity of the lime will also bring about precipitation of the soluble chromium salts. Vegetable tanning materials and soluble proteins from the soak water will interact to form a sludge which can be removed later by settling. Coagulation of tannery wastes, therefore, is almost a natural process due to the reactivity of the component parts from the various departments of the tannery.

The quantity of materials used in the tannery does not result in a completely balanced sludge. In most cases there is an excess of beamhouse wastes and the tannery discharge problem is lime and sulfide. In other cases, the balance may present entirely different problems.

c) Settling. By the use of lagoons, the coagulated and mutually interacted portions of the tannery effluent will settle and a sludge will form in the bottom of the lagoon. The water may be passed through several lagoons, and finally it will discharge directly with an acceptable degree of purity. Trickling filters or aeration systems may be employed to aid in the bacterial decomposition of the tannery effluent, in the clarification of the spent liquor, and in a decrease in the BOD.

d) Oxidation of sulfide. The disposal of sulfides must be by chemical means. No beamhouse operation can be conducted in such a way as to discharge the low levels of sulfide required. Treatment of the wastes is needed. The standard method used is by the air oxidation with the use of a manganese sulfate catalyst. This system can be done using air bubbling of the solution in a tank, or by use of aerators. By either system, the sulfide can be quantitatively removed in approximately 4 hours. The sulfide free wastes can then be acidified to neutral pH if desired.

TABLE 4

Promulgated Guidelines for July 1979
Best Available Treatment Technology Economically Achievable

Pollutant or Pollutant Property				Subcategory*		
	1	2	3	4	5	6
				(Parameter kg/kkg)**		
BOD$_5$	0.61	0.65	0.47	0.23	0.40	0.44
COD	5.8	6.2	4.5	2.2	3.8	4.2
TSS	0.70	0.74	0.54	0.26	0.46	0.50
Oil and Grease	0.26	0.28	0.20	0.10	0.17	0.19
Total Chromium	0.015	0.015	0.011	0.005	0.0096	0.010
TKN	0.66	0.69	0.51	0.24	0.43	0.47
Ammonia	0.22	0.23	0.17	0.081	0.14	0.16
Phenol	0.0043	0.0046	0.0034	0.0016	0.0029	0.0031
Sulfide (mg/l)	0.0	0.0	0.0	0.0	0.0	0.0
pH	Within the range of 6.0 to 9.0 at all times					

The above values are the permissible maximum average of daily values for any period of thirty consecutive days. Maximums for any one day may be double the value shown for BOD$_5$, Chromium, Oil and Grease and Suspended Solids.

*Categories

Category 1. Hair Pulp, Chrome Tan. This is the present method of producing most cattlehide leather. The system results in high BOD, high sulfide, high suspended solids, some chromium, some oil and grease, low pH and some fecal coliform.

Category 2. Hair Save, Chrome Tan. This method is used by some cattlehide tanneries. The system results in an effluent similar to category 1, but somewhat lower BOD, sulfide and suspended solids. Otherwise the effluent is similar to category 1.

Category 3. Hair Save, Vegetable Tan. This method is used in processing cattlehides for sole leather. The system results in an effluent similar to category 2, with respect to sulfide. BOD and suspended solids are usually higher than category 2 tanneries. There is no chromium used in this process.

Category 4. Hair Previously Removed, Previously Tanned. Several types of skins fall into this group; splits, Indian buffalo, some pigskins, etc. The effluent should have relatively low BOD, suspended solids and low pH. No chromium or sulfide is used in the process.

Category 5. Hair Previously Removed, Chrome Tanning. This includes pickled hides and skins. Many sheep skin tanneries fall into this group. The effluent should have low BOD, suspended solids, low pH and high chromium. No sulfide is used in the process.

Category 6. Hair Pulp, Chrome Tanning, Without Finishing. This is similar to category 1, with regard to BOD, chromium and sulfide. Suspended solids and oil and grease may be somewhat lower than category 1.

**kg per 1000 kg of hides processed

Table taken from U.S. Environmental Protection Agency, Proposed Effluent Limitations Guidelines, New Source Performance Standards, Washington, D.C., July 1979. .

Neutralization of pH 5.0 to 6.0 will result in the precipitation of large quantities of soluble proteins. This system has been proposed as a means of decreasing the BOD. A recoverable protein precipitate from this approach may yield a fertilizer or an animal feed.

e) Removal of chromium. Chromium may be removed by mixing with the alkaline beamhouse wastes. This system can result in the removal of chromium to nearly the levels desired by the EPA, providing an excellent settling system is available. In most cases, however, the chromium remains as a colloidal suspension and the quantity of chromium discharge is more than permitted.

Some tanneries have taken the step of recycling the chrome tanning liquors. In drum tannages, where solutions are relatively concentrated, recycling can be employed directly. There will be a build up of excess spent chromium liquor. To deal with this, the solution can be neutralized to pH 8.0 or 9.0 and the chromium will precipitate. The solution is then decanted and the precipitated chromium sludge disposed of separately. The sludge may be acidified and reused or disposed of with solid wastes.

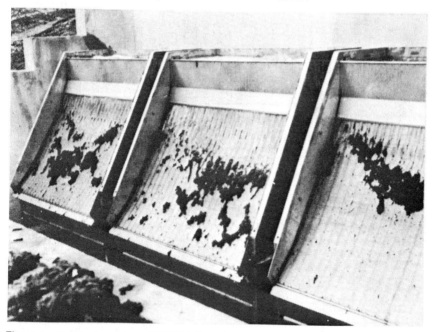

Figure 1   Solids screening. (HYDRASIEVE) Wedge wire screens such as this are being employed by many tanneries for the removal of rough solids. Buffing dust, hair, shavings, etc., can be removed by this system. *(Photo courtesy of The Bauer Brothers Company).*

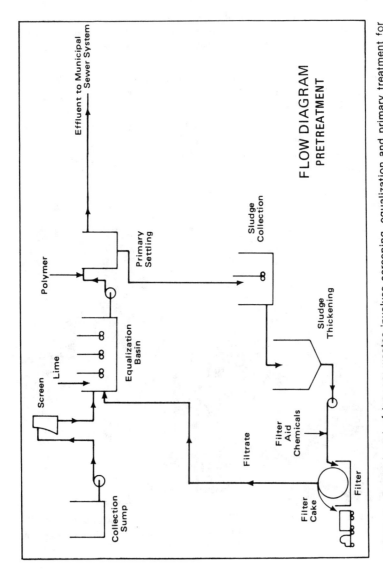

Figure 2 Pretreatment of tannery wastes involves screening, equalization and primary treatment for solids removal. Chromium, tanning, recycling and sulfide oxidation may also be necessary. *(Taken from Development Document for Effluent Limitations Guidelines and Standards of Performance, Prepared by Stanley Consultants, Inc.).*

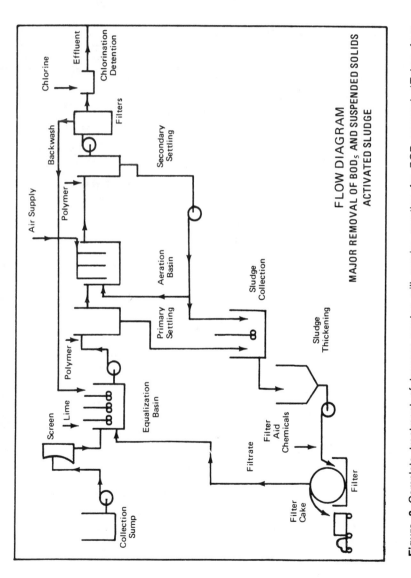

Figure 3 Complete treatment of tannery wastes will require aeration for BOD removal. (*Taken from Stanley Consultants, Inc. report—See Figure 1*).

Figure 4   Sludge dewatering filter press. Coagulated and settled primary wastes from pretreatment systems of primary and secondary systems may be dewatered by means of a filter press. Dewatering is necessary to achieve a high solids content to make handling easier and decrease costs. Solids content of the sludge can be as high as 50% to 60% by this system. *(Photo courtesy of PolyFilters, Inc.)*

Specific effluent treatment systems

The designs given below of certain specific effluent treatment systems for tanneries are only to serve as a guide and to familiarize the reader with the general nature of the systems.

Under a study of tannery wastes by the New England Interstate Water Pollution Control Commission, detailed investigations of a pigskin tannery and a cattle hide tannery were made covering the analysis of the effluent from the various departments and the quantities of chemicals used. This was calculated on the basis of the quantity of hides soaked. The study, as with many previous ones, indicated that the bulk of the pollution load is contributed in the early stages of leather manufacture. In the cattle hide tannery, 70–80 pounds of BOD are produced from each thousand pounds of wet salted cattle hide. Most of this pollution load (85%) is produced in the same three processes: 20% in soak, 52% in unhairing, 13% in deliming and bating. Protein and other substances extracted from the hide are estimated to produce 50–70% of the total BOD of the load, and process chemicals are estimated to produce 30–50%. Of these, the sulfide ion is the most significant, producing approximately 10% or more of the total BOD load, depending upon the type of unhairing system used. The discharge of effluent from a tannery should average between eight and twelve thousand gallons per thousand pounds of wet salted hide processed.

Based on this information, two effluent treatment systems are recommended—one for a tannery with large land area, and the other for a tannery with small land area or with minimum space available.

For a tannery with a large area (Figure 5). The entire quantity of tannery effluent, after preliminary screening, is brought into a settling lagoon capable of taking several days' effluent. At least two settling lagoons are necessary. The first lagoon is filled and the effluent allowed to settle for several days (one to seven days is recommended). The overflow from the settling lagoon then goes into an equalizing lagoon for neutralization to pH 9 by the addition of carbon dioxide. The carbon dioxide used is from the spent flue gases. Further sedimentation results, and final passage of the effluent through the trickling filter, prior to discharge into the stream, takes place. The sludge can be removed from the equalizing lagoon and pumped into a storage lagoon where it can accumulate until it is removed, sold as fertilizer, or buried. This treatment requires a large area of land due to the tremendous volumes of water being processed and the large quantities of sludge.

For a tannery with minimum land area (Figure 6). The settling tank should be kept as small as possible. Also, pumping and forced filtration should be employed. The water from the soak, unhairing, deliming, and bating should be equalized together as strong liquors, and the sludge should be recovered for fertilizer or fill material. From the other sections of the tannery the acid, quantities of the pickle, the water or brine portion from the degreasing operation, and the water portion from the coloring, fatliquoring, and finishing operations are brought together into a sump and used to neutralize the effluent from the beamhouse equalization tank. The resulting solution should be low in solids and have an alkaline pH. The addition of carbon dioxide will precipitate calcium carbonate and will remove additional suspended organic materials. The clarified liquor, after settling, can be passed through a trickling filter, or aerated; after an additional filtering it can be discharged into the stream. This system should result in a very clear and effective treatment with minimum land use. A relatively high investment in equipment is required.

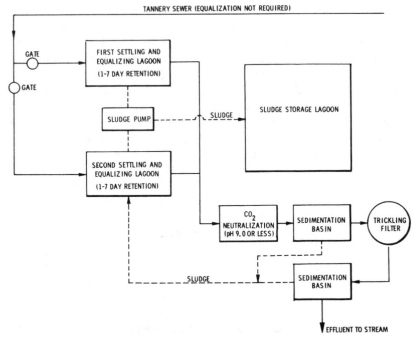

Figure 5   Plan for an effluent plant for a tannery having a large land area.

TRENDS IN THE INDUSTRY IN EFFLUENT CONTROL

Sulfide removal

There have been efforts in the development of nonsulfide unhairing systems as described earlier. These have received limited commercial acceptance. Others have attempted to decrease the quantity of sulfide use in the unhairing system. This in itself is not successful since even at the lowest levels of sulfide, the amount of sulfides present in the effluent is too high to meet the requirements. It is necessary, therefore, the treat the sulfide for sulfide removal in both the direct and the indirect discharge systems.

Sulfide removal in the beamhouse follows two distinct courses.

(1) The most common method of removing sulfide was decribed earlier: the oxidation of the sulfide with a manganese catalyst and the addition of air. This has been studied by a number of different workers and is in practical use in many tanneries in the United States.

(2) An alternate method that has proved successful is based on the removal of the sulfide as hydrogen sulfide by acidification, collecting the sulfide in an alkaline bath for recycling. This system can be effective in a very large tannery and is an example of what can be done by proper engineering and chemical control.

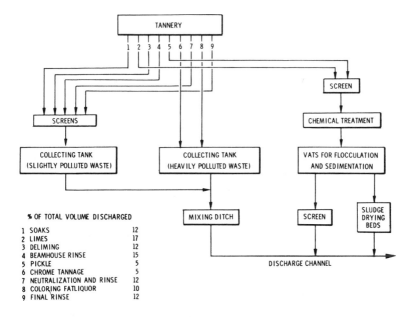

Figure 6   Plan for an effluent plant for a tannery having a small land area. The separation of concentrated waste from dilute waste, neutralization, and co-precipitation of selected wastes is the basis of this system.

The following system has been operated successfully by Prime Tanning Co., Blueside Division in St. Joseph, MO. The tannery receives fresh and cured cattle hides. The hides are soaked, unhaired, bated, pickled, and tanned through chrome tanning. No further operations are conducted.

Beamhouse wastes are sent to the clarifier for settling. Chrome tanning wastes are collected and precipitated for recycling. The overflow from the clarifier, being at a high pH (10–12), contains small quantities of lime and suspended matter. The clarified overflow contains sulfides.

Clarified waste is acidified with sulfuric acid in the acid scrubber where it flows from plate to plate from the top down to the bottom. There is a counterflow of air from the column to carry away the hydrogen sulfide generated. The hydrogen sulfide is carried with the air into a second scrubbing column where a spray of sodium hydroxide flows over the packed column and scrubs the hydrogen sulfide.

Acid Scrubber:

$$Na_2S + H_2SO_4 \quad \xrightarrow{(pH < 5)} \quad Na_2SO_4 + H_2S \text{ — to alkaline scrubber}$$

Alkaline Scrubber:

$$H_2S + NaOH \quad \xrightarrow{(pH > 10)} \quad NaSH \text{ — to recycle}$$

In the acid scrubber, sulfuric acid is added into the clarified alkaline waste, lowering the pH. The pH is dropped to below 5.5. The sulfide and sulfydrate in the waste are converted to hydrogen sulfide and are carried by the air and recovered in the scrubbing column. The waste at pH 5.5 or so can be adjusted to 6.0–9.0 and discharged into the municipal sewer system. The final effluent usually contains essentially no chromium, no sulfide, and only a small amount of suspended solids and BOD. The sulfide accumulates in the alkaline tank and is recycled for use in the beamhouse.

The majority of the waste removed from the primary clarifier is sludge. This sludge is dewatered in a filter press and the waste sludge is disposed of by land application. This system is quite effective and is particularly useful where there is sufficient agricultural land that can accept the sludge.

The importance of chemical control when working with sulfide bearing wastes cannot be underestimated. Such a system as described above must be properly designed and operated to avoid the dangerous release of toxic hydrogen sulfide.

Chromium removal

There has been considerable emphasis on the removal of chromium from tannery wastes. Trivalent chromium is not objectionable in trace quantities. Recently, regulations have been modified so that land disposal of alkaline waste sludges containing precipitated chromium is acceptable. Tannery sludge and tannery waste shavings are acceptable as raw materials for fertilizer and modification into fertilizer products. Restrictions on indirect

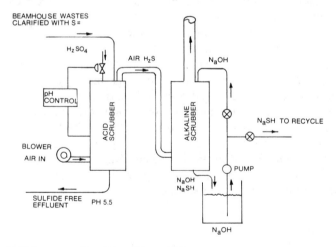

Figure 7   Sulfide removal by the acid scrubbing system.

discharges of chromium in a municipal system are based primarily on the concern over the incineration of sludges and the formation of hexavalent chromium under these conditions.

The removal of chromium may be by direct recycling, recycling by precipitation with an alkali, or by co-precipitation of the chromium with beamhouse wastes.

Liritan system

The liritan system was described in Chapter 9. This is a so-called "no effluent system" in that the vegetable tanning materials are recycled. It is recognized as an acceptable means of decreasing the total quantities of suspended solids and BOD in vegetable tanning. It is not applicable to retan systems.

Blue side systems

With the shifts in the early stages of tanning production to the blue side operations, tanneries have been set up in rural areas having dry, warm weather for most of the year. In these areas there are no rivers available to take any kind of treated effluent. The municipalities are relatively small and unable to handle any large quantities of waste generated by a tannery of even small size.

By recycling chrome liquors and recycling all solutions as much as possible, the total volume of water can be decreased to a fraction of that listed in the early parts of this chapter. Several ideas have been suggested such as using spent chrome tanned liquors for makeup of pickle stock. Spent bate water may be used for soak water. The wastes can be run into open lagoons for solar evaporation. Evaporation of the concentrated waste removes the water and de-waters the sludge. The sludge containing lime and protein material can be applied as a fertilizer. It is possible to completely eliminate any discharge to a river or municipal sewage disposal system by this type of planning.

Effluent treatment for a blue side tannery in an isolated hot dry climate

Blue side tanneries in the United States are best located near a source of supply of fresh hides. In the hide producing areas of the Southwestern United States, there are also large open land areas and a hot dry climate. The possible use of recycling of liquors for maximum water conservation is illustrated in this example. The requirements for disposal by evaporation are also illustrated.

We can assume the following flows (50,000 lbs hide/day)

soak	50,000 gal/day
unhairing (hair burn)	50,000 gal/day
lime	100,000 gal/day
bate	500,000 gal/day
chrome tan	recycle, no discharge

The flow gallonages may be higher or lower in each process in accordance with the tanner's methodology.

All wastes are segregated

(1) Soak wastes go directly to the primary clarifier.

(2) Unhairing wastes, heavy with sulfides, suspended solids, fats, and dissolved organic matter, are sent for oxidation using manganese catalyst and air. The sulfide-free unhairing wastes go to the primary clarifier.

(3) Spent lime solutions may be used for unhairing. To do this, the lime slurry is added to the soaked drained hides and sulfides are added. The lime wastes not recycled for unhairing may be sent to the primary clarifier, since the sulfide content is low.

(4) Bate solutions contain ammonium salts and proteins both dissolved and suspended. The bate is sent to the primary clarifier.

(5) In the primary clarifier, the solids are removed and pumped to sand beds for drying. Assuming 10% solids in the sludge, we would expect about 10,000 gal/day of sludge to be air dried. Assuming no seeping into the soil, about one to four acres would be required for the sludge drying area, depending upon the weather.

(6) In the secondary lagoon the BOD will be removed by microbiological action as is normal. By using a large lagoon, the system employs extended aeration which does not result in secondary sludge. An alternate method is to employ a secondary clarifier.

(7) The treated wastes are neutral in the pH and contain some suspended solids and some BOD. There is no chromium in the waste. The solution can then be recycled back into plant processes of soak, liming, or bating, to whatever extent the tanner wishes. The excess solution can be run into an effluent pond for disposal by evaporation. Approximately one acre of pond (depending upon local weather conditions) will be required for each 1,000 gallon discharge per day.

The disposal of sludge and possibly the treated waste water in agricultural application may also be an advantage.

Pretreatment system for an indirect discharge

For the chrome tanning tannery with a beamhouse, discharge into a publicly owned treatment works (POTW) is usually desirable. Pretreatment to eliminate essentially all of the chromium and sulfide is required. The decrease in suspended solids is also usually required.

As shown in the figure 9, the beamhouse wastes containing the sulfides are all collected and treated by sulfide oxidation using a manganese sulfate catalyst and air. This will remove the sulfide to only a few parts per million in a batch process.

The treated beamhouse wastes are then pumped to the clarifier at a controlled rate where they are mixed with the wastes from the rest of the tannery. The interaction of the two waste streams will result in an alkaline mixture of pH 10-12 in which co-precipitation will occur. Chromium, tannins, dyes, and spent fatliquors will all be precipitated to a greater or lesser extent.

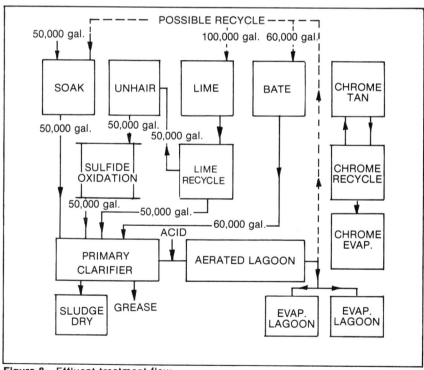

Figure 8   Effluent treatment flow.

The sludge from the clarifier is pumped by pump B (a positive displacement pump) to the filter press for dewatering. The dewatered sludge should be 30–50% solids, containing precipitated chromium, hair and leather fibers, lime, and miscellaneous tannery wastes. This should be acceptable for landfill. The water from the filter press is returned to the clarifier.

The clarified water from the clarifier is essentially free of sulfides and chromium, and has a pH of about 10–12. Most public sewer systems require a pH of from 5–9. A slight adjustment in pH may be needed to bring the pH to the desired range.

LOOKING AHEAD

Most of the tanneries in the United States have now consolidated to relatively large units and have put forth sufficient effort to solve the major portion of the pollution control problems. Continued upgrading of the effluent, both direct and indirect, will of course be necessary. Many tanners now feel that the pollution problems are under control, and they can concentrate their technical efforts on making more competitive leathers.

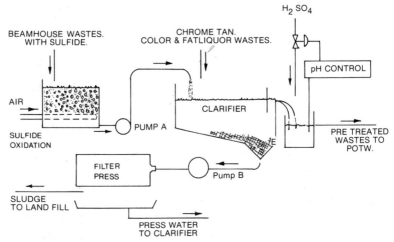

Figure 9   Pretreatment system for an indirect discharger.

REFERENCES

Eye, J. D., and Graif, S. P., *J. Am. Leather Chemists' Assoc.*, **62**, 194 (1967).
Pepper, K. W., *J. Am. Leather Chemists' Assoc.*, **61**, 570 (1966).
Reuning, H. T., *J. Am. Leather Chemists' Assoc.*, **42**, 573 (1947).
Rudolfs, W., "Industrial Wastes," ACS Monograph 118, New York, Reinhold, 1953.
Tannery Wastes, New England Interstate Water Pollution Control Commission, Wesleyan University (1958).

BIBLIOGRAPHY

Protection of the Environment
    Title 40, Chapter 1, Subchapter N, Part 425, Leather Tanning and Finishing Industry Point Source Category, Federal Register Volume 39, No. 69, Tuesday, 9 April, 1974.
Development Document for Proposed Effluent Limitations Guidelines and New Source Performance Standards for the Leather Tanning and Finishing Point Source Category, U.S. Environmental Protection Agency, November 1973.
Tannery Wastes—Pollution Sources and Methods of Treatment. New England Interstate Water Pollution Control Commission, June 1958.
Economic Analysis of Proposed Effluent Guidelines—Leather Tanning and Finishing Industry, U.S. Environmental Protection Agency, October 1973.
Pretreatment of Pollutants Introduced into Publically Owned Treatment Works, Washington, D.C., 1973.
Protection of the Environment—Federal Register

LEATHER TESTING
AND ANALYSIS

19

Leather is, by nature, an inconsistant material. The hides and skins vary from one to another according to the breed, environment and experiences of the animal. At the time of skinning, mechanical damage may occur. Curing the skins may differ greatly and result in loss of strength or grain character. The evaluation of leather quality is, as a result, a matter of subjective judgment and opinion. In modern commerce it is essential that quality be defined in objective terms and that buyer and seller agree on methods of determining the suitability of a product for the intended use.

The problem of testing and evaluating leather has been a subject of great concern to leather technologists throughout the world. The tests now in use were developed by committees of technical specialists with cooperative evaluation of the proposed tests by the members. This work has resulted in a number of tests that have been accepted by the industry in establishing standards and specifications. Since leather is so diversified in its desired properties, complete specifications are not the normal industry practice except for military and a few other uses. More often tests of a particular property are used as referee data to settle disputes. For quality control the leather technician may develop quick tests that are sufficient for the purpose. However, if the data are to be used to meet a specification, settle a customer's claim or guarantee a particular property, the published method should be followed exactly and referenced in the report.

The tests in common commercial practice can be divided into three classes. These are:

Mechanical tests of strength—stretch, stitch, tear, etc.

Moisture related tests—water absorbtion, water vapor transmission, resistance to water penetration, etc;

Chemical Analysis—oil content, tannin, hide substance, etc.

The number of tests is far too great to include in this book a discussion of more than few of the most common and most representative. More complete details of tests and their significance are available from the participating organizations:

American Society for Testing Materials
American Leather Chemists Association
International Organization for Standardization
International Union of Leather Trades Chemists

SAMPLING OF LEATHER

Because leather is such a variable product it is essential that the samples for testing be representative of the test lot. Samples should be taken in an agreed location on the hide or skin. It is also important that the number of samples be sufficient to result in statistically meaningful data. The location of the direction of the sample in the hide should be indicated for physical testing.

Physical tests:

Tensile strength. Tensile tests are done with dumbell shaped samples cut with dies that meet the design indicated in the method. The cross section of the sample is measured and calculated. Samples are broken and the breaking strength per unit of cross sectional area calculated. The results are averaged. In some cases, if agreed, a low sample may be discarded as not being representative.

The importance of physical strength in leather is obvious. This strength may be critical in tensile strength as in belting or mechanical leathers or it may a matter of stitch tear or eyelet strength. In the tensile test elongation of the leather at a particular load may be critical.

Elongation is measured by noting the stretch of the leather between two marks on the test sample. By use of divider points of a compass the elongation can easily be measured. The results are given in percent. The load on the leather may be specified in the measurement of elongation of a particular leather.

Grab method. For light leathers the dumbell shaped sample may not give reliable results because of the edge effect on the thin sample. In this case the sample is cut wide enough so that the edges do not enter into the strength test. The sample taken is much wider than the jaws of the test machine to eliminate the edge effect. The jaws of the test machine are the determining factor for the width the test.

Stitch tear. The stitch tear is applicable to shoe upper leather and light leathers. In this test a wire loop is pulled through two prepunched holes in the leather. This simple test is effective and will indicate leather weakness in garment and glove leathers.

Slot tear. This test measures the strength of the leather in resistance to tearing from a precut slot. The test is good on shoe upper leather.

Single hole tear test. This test is particularly good on thick samples. The test is similar to the stitch tear test but eliminates the cutting effect of the wire.

Moisture Related Tests

Leather may be resistant to water but the industry is reluctant to claim it is "waterproof." The leather may pass water vapor through but resist liquid water penetration. The leather will also absorb some liquid water and yet be water resistant. In order to clarify these factors several tests have been developed. The most frequently used are the moisture vapor penetration, the water absorption, and the dynamic water penetration. These tests are all significant for shoe upper leather.

Moisture vapor penetration. This measurement is made by sealing a sample of leather over the top of a weighted dish containing a desicant. The test unit is placed in a controlled atmosphere chamber (high humidity). The weight gain of the unit is measured as a function of time to obtain a penetration rate per unit area. The moisture vapor penetration is an important factor in the comfort of shoes.

Moisture absorption. Weighed samples are soaked in water, either in a static basin or in a tumbling container, for a specified period of time. The samples are removed, blotted, and weighed. The percentage of weight gain is reported.

Dynamic moisture penetration. Leathers that have been treated to resist water when made into shoe uppers are tested by the dynamic moisture penetration machine. In this test the leather is folded into a trough. The trough is placed in water and the leather flexed in a manner similar to the action of the foot during walking. Water penetration is indicated by completion of an electrical circuit. Fifteen thousand flexes is considered a Pass by the U.S. Army. This is equivalant to walking in water for more than ten kilometers.

CHEMICAL ANALYSIS OF LEATHER

Chemical analysis of leather is as reliable a measure of leather quality as any physical test. The analytical test methods are relatively simple and can be done with only a minimum of laboratory equipment. The significance of the analysis will depend on the type of leather and its intended use.

The leather to be analyzed must be a composite sample from the designated sampling area from a enough pieces to produce significant results.

Data taken from improper samples will result in errors that will lead to wrong conclusions. The sample pieces are ground to a fine powder and mixed to obtain a uniform sample. Duplicate analyses should be done to assure accuracy. If the duplicate data do not agree within the accuracy of the procedure the analysis should be repeated.

The samples for chemical analysis should be weighed all at the same sitting. Leather is hydroscopic and, therefore, the moisture content of the leather will change from day to day. The analyst should determine the total weight of samples needed, then grind an excess amount of leather. This leather is then well mixed and all samples weighed.

Moisture content. The moisture content is determined by weighing the sample in a weighed crucible and drying the sample, usually overnight. From the weight loss and the initial weight of the sample, the moisture content is calculated. Leather analyses are reported on a moisture free basis. The moisture content is used to convert the ''as received'' data to the ''moisture free basis.''

Ash content. The ash content may be done on the dried sample from the moisture determination by simply placing the crucible and dried sample from the moisture determination into a laboratory furnace at the desired temperature for the time needed.

Extractable matter. The oil content of leather is of great importance. Fat-liquoring oils are needed to lubricate the fibers for softness and strength. An excess of oil or grease will prevent proper adhesion of finishes or adhesives used in shoe manufacturing. To determine the oil content of leather, a sample, as previously prepared, is weighed into a porous extraction thimble and extracted with a Soxhlet extractor. The solvent containing the oil is evaporated and the remaining oil weighed. The results are reported of a moisture free basis. There is a difference of opinion among leather chemists about the best solvent. Some prefer hexane, others prefer dichloromethane or chloroform. The method used is referenced in the chemist's report. The solvent to be used is specified as part of the ''standard method.'' The difference in the extraction is usually negligible.

Hide substance. To measure the amount of hide substance in leather, a Kjeldahl analysis is used. The sample of leather is weighed then digested with sulfuric acid, sodium sulfate and a catalyst. The digestion reaction converts the protein to carbon dioxide and ammonia. The solution is cooled, neutralized with sodium hyroxide to strongly alkaline, and the ammonia distilled into an acid receiving bath. The ammonium hydroxide is titrated and the weight of the protein calculated. Because in most leathers the only significant nitrogen is in the hide protein, the weight of the nitrogen multiplied by a factor (5.62) will give the weight of the hide substance.

Chromium oxide. It is customary to report the chromium in leather as chromium oxide, (Cr_2O_3). This does not mean that the chromium is present as the oxide. In chrome tanned leather the chromium is present in a combined form with the hide protein. The result of the analysis is often reported as Cr_2O_3 on a hide substance basis. This is more meaningful for leathers that contain syntans or vegetable tannins.

The chromium analysis can be done on the ash sample from above or it can be done on the leather directly. The weighed sample is digested with sulfuric, nitric and perchloric acids to oxidize the chromium of the hexavalent state. The solution is diluted and the chromium determined by titration with a reducing agent.

CAUTION: This is a very dangerous analysis. The hot perchloric acid digestion gives off strong toxic fumes so a good fume hood is needed. Also the hot solution can explode by the reaction of the perchloric acid with the organic matter present. Help and training by a chemist familiar with this analysis is *strongly* advised.

Nondetermined substances. Included in the analysis of leather are vegetable tannins, syntans, and other organic substances. Leather analysis will determine the moisture, ash, oils, and hide substance. These are all direct analyses. The rest of the leather is a mixture of substances that cannot be directly determined by methods presently available to leather technologist. In order to estimate these materials, a calculation can be made; the moisture, ash, extractables, and hide substance percentages are subtracted from 100 and the remainder is the nondetermined substance. This nondetermined substance is assumed to be tannins in vegetable tanned leathers. From this the nondetermined substances divided by the hide substances, calculated as a percentage is reported as the degree of tannage. This is only significant in vegetable tanned leathers.

Water soluble matter in vegetable tanned leather. This type of leather is often filled with materials that are not firmly bound to the leather fibers. To determine the water soluble matter the leather, after extraction of the oils and fats as given above, is extracted with water at a specified temperature (35° C) for about three hours. The water is then evaporated and the residue weighed. This is only significant for vegetable tanned leathers.

pH. The pH of leather is important because it is an indication of the stability of the leather over a long period of time. To measure the pH of leather a ground sample is soaked in distilled water for a given period of time and the pH of the solution determined with a glass electrode pH meter.

Shrink temperature. Chrome tanned leather, when properly tanned is very resistant to hot water. The measurement of the shrink temperature is very important for quality control in the tannery and is a quick test of the tannage effectivness in leather. The test can be made on leather at any stage of production, during or after tanning, and any time after that. The leather need not be equilibrated for moisture content. The sample is cut as a strip about 1 cm wide and about 5–8 cm long. The sample is clamped between the two jaws of the shrink meter, the assembly is put into water or a water-glycol solution and the temperature raised gradually. When the shrink temperature is reached the leather will decrease in length. This decrease is shown by a mechanical system that indicates small changes in the length of the sample. Well-tanned chrome leathers will shrink at over 95° C or more. Many tanners consider any shrink temperature lower than 100° C unsatisfactory.

Shrink temperature is not of commercial significance in vegetable leathers.

To illustrate the relative physical test and chemical analyses of some different leathers, listed below are the partial test specifications for four different leathers. These requirements should not be taken as standards for any other leather. The purpose of the data is to illustrate how different these leathers are.

Military Boot Leather

Thickness	4.5–6 oz. (1.75–2.4 mm)	Chloroform Soluble	6.0–16.0%
Slot Tear	35 lb. 80% pass	Cr_2O_3 on Hide Subst.	4.0–7.5%
Shrink Temp.	97° C	Nondetermined Material	20% Max.
pH	3.0–4.0	Ash on Moisture Free	10% Max

Vegetable Tanned Strap Leather

Thickness and Strength on .5″ sample	2–3.5 oz. 3.75–5.5 oz. > 5.75 ox.	30 lb. 65 lb. 110 lb.	Elongation 20% Max.
Stitch Tear	30 lb. 80%		pH 3.0–5.0
Chloroform soluble	9.0–15%		Total Ash 5.0% Max.
Water Solubles	18.0% Max		

Work Glove Leather

Thickness and Stitch Tear	< 2.5 oz. > 2.5 oz.	17 lb. 30 lb.	Elongation 20–60% at 25 lbs.
Chloroform soluble	25% Max.		Cr_2O_3 on Moisture free 3.0 Min.
pH	3.3 Min.	Total Ash 9.0 Max.	

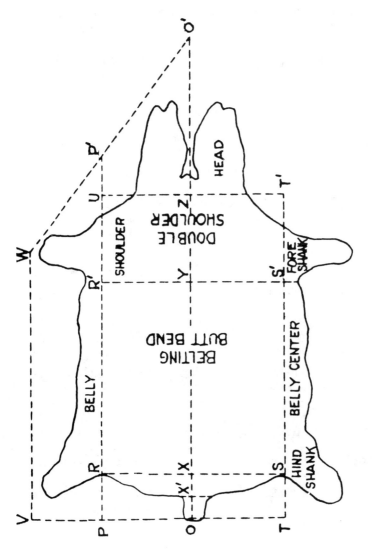

FIGURE 1–LOCATION OF HIDE PARTS

The enclosed figure is part of the Definitions and Terms section. This was omitted (by error) from the third edition.

DEFINITIONS OF TERMS USED IN THE LEATHER INDUSTRY

The definitions of terms listed are from Federal Test Method Std. No. 311, January 15, 1969

SCOPE. The definitions of general terms include some of those encountered by personnel concerned with specifications and procurement of hides, leather and leather products by the Federal Government. Terms adequately defined by unabridged dictionaries are not generally included.

GENERAL TERMS

Alligator-grained leather. Leather of various types, such as calf, sheep, or cattle-hide embossed to resemble the grain of alligator hide.

Apron leather. Any one of several varieties of leather used in connection with textile machinery and blacksmith aprons. Comber and Gill Box Apron leather is soft, mellow, tough leather, tanned from steer hides, heavily stuffed and boarded or otherwise softened. Rub Roll apron leather is a flexible but firm, dry, strong leather.

Aspergillus niger. One of the most common mold growths found on leather, usually greenish or blackish in color.

Back. A crop with the head trimmed off behind the horn holes. (OZUP in Figure 1.)

Bag leather (Also know as case leather). A general term for leathers used in traveling bags and suitcases. It does not include the light leathers employed for women's fancy handbags. The staple material for bag and case leather at present is leather made from the hides of animals of the bovine species, but heavy sealskins and goatskins are also used.

Barkometer. A hydrometer used for determining the specific gravity of tanning solutions. A specific gravity of 1.000 is equivalent to $0°$ barkometer (Bk), and each additional degree Bk is equivalent to an increase of 0.001 in specific gravity.

Baseball leather. Leather used for covers of baseballs. The better grades of balls have covers of alum-tanned horsehide front leather. Some cheaper grades are made of kip and sheepskins.

Bate. To treat unhaired hides or skins with a warm aqueous solution of enzymes in order to remove certain undesirable nitrogeneous constituents.

Beam. A convex wooden slab sloping downwards from about waist heights, over which a hide is placed for trimming off excess flesh and ragged edges and for unhairing by hand.

Belting butt. A double back with the tail cut off at the butt line. (RUT'S in Figure 1).

Belting butt bend. A double bend with the tail cut off at the butt line. (RR'S in Figure 1).

325

Belly. That part of the hide below the belly line. (VWP'P in Figure 1). For steerhide leather, the belly line (RU) passes through a point at or above the top of the rear break. For cowhide leather, the belly line passes through a point at or above the top of the front break and a point not more than 2-1/2 inches below the top of the rear break.

Bend. A back with the shoulder cut off at right angles to the backbone line at the break of the fore flank. (OYR'P in Figure 1).

Biff. To beat a salted hide that has been placed on a rack, in order to shake loose salt from the hair.

Bisulfiting. The treatment of hot solutions of vegetable tanning extracts with sodium bisulfite in order to increase their solubility and rate of take-up by hides.

Bleaching. (1) The process of removing oxidized tannins and insoluble materials from the surface layers of leather, particularly sole leather, in order to prevent crackiness of the grain and to improve color. It is performed by dipping the leather in a weak alkaline solution to render the tannin readily soluble, dipping in water, neutralizing in weak acid solution, and washing. (2) The process of lightening the color of chrome leather by treating with synthetic tannins or precipitating white pigment in the surface of the leather.

Bleeding. The transfer of materials exuded from leather to other material that comes into contact with it. It is usually designated as staining.

Blocking. The adhesion between touching layers of similar or dissimilar material, such as occurs under moderate pressures during storage or use.

Bloom. A light-colored deposit of ellagic acid appearing on the grain surface of leather tanned with certain pyrogallol tannins, such as myrabolans, valonia, and dividivi. The appearance may be objectionable for some purposes, but bloom does not significantly affect the other physical properties of the leather.

Blue. Usually in the phrase "in the blue". Applied to hides or skins that have been chrome-tanned but not dyed or fat liquored.

Boarded leather. Leather on which the grain has been accentuated by folding the grain side in and working the leather back and forth. Hand boarding is done with a curved cork board attached to the worker's arm and rolled over the folded skin.

Boardy. Adjective applied to stiff, inflexible leather.

Break. (1) Heavy leather—The places, in the area where the fore shank and hind shank join the body of the hide, where the texture of the leather changes quite sharply from the firm, close weave of the bend to a loose, open texture. (2) Shoe upper leather—The superficial wrinkling formed when the leather is bent, grain inward, with a radius of curviture like that formed at the vamp of a shoe in walking. Adjectives commonly used to describe this characteristic are tight, loose, coarse, and pipey.

Brining. A process of curing hides by washing and soaking in a concentrated salt solution.

Brush coloring. The application of dyestuffs by brushing.

Buck sides. Cattlehide shoe upper leather finished to resemble buckskin.

Buckskin. Leather from deer and elk skins; used for shoes, gloves and clothing. Only the outer cut of the skin from which the surface grain has been removed may be correctly defined as "genuine buckskin". Leather finished from the split or under cut of deerskin must be described as "split buckskin".

Buffalo. Buffalo leather is made from the hides of domesticated water buffalo of the Far East, not the American bison.

Buffing. (1) Removing minor blemishes from the grain with a knife or abrasive. (See Snuffed top grain.) (2) Producing a velvet surface on leather, usually with an emery wheel. (3) Buffing leather is a light cut of the grain portion used for bookbindings, pocketbooks, etc., but not for upholstery.

Bullhides. Hides from bulls are characterized by thick and rough head, neck and shoulders, and by coarse flanks. Bullhides are often poor in quality and heavy, ranging from 60 pounds up.

Butcher cuts. Damage to hides caused by improper removal from the animal. Damage is usually in the form of cuts or furrows on the flesh side.

Butt. That part of the hide or skin covering the rump or hind part of the animal.

Cabretta. Skin of Brazilian hair sheep used principally for glove leathers. Term probably derived from Spanish "Cabrito", or similar Portuguese or Italian word. (cf. Cape).

Calf leather. Leather made from the skins of young cattle from a few days up to a few months old, the skins weighing up to 15 pounds. Calf leather is finer grained, lighter in weight and more supple than cowhide or kip leather.

Cape (skin or leather). Skin of South African hair sheep. Fine-grain leather, superior to wool sheep for gloves and garments. Loosely applied to all hair sheep, but should be qualified to show origin, if other than South African. (Uncertain whether term is derived from "Caper" (Goat) or from "Cape Town".

Carding leather. A type of side leather used on the cards of textile machinery.

Carpincho leather. Leather from the skin of the carpincho, a large South American rodent. The skin is used in making glove leather, usually chrome tanned and washable. In the glove-leather trade, carpincho is classified as a pigskin. It resembles pigskin in appearance, because of the occurrence of bristle holes in straight-line groups, usually with 5 but may vary between 4 to 7 holes in a group.

Case leather (Also known as bag leather). A general term for leathers used in travelling bags and suitcases. It does not include the light leathers employed for women's fancy handbags. The staple material for bag and case leather at present is leather made from the hides of animals of the bovine species, but heavy sealskins and goatskins are also used.

Chamois leather. A soft pliable absorbent oil tanned leather which is recognized in this country and abroad as being made from sheepskin, from which the outer or grain side has been split prior to tanning, known technically as a flesher.

Chrome retan. Term applied to leather tanned first with chromium salts, then retanned with vegetable extracts.

Chrome retannage. Retannage with chromium salts.

Chrome tannage. Tannage of leather with chromium compounds. Chrome tanned leather is often distinguished from other kinds by its greenish color, particularly of a cut edge.

Cockle. A hard warty growth on sheepskin.

Collagen. The principal fibrous protein in the corium or derma layer of a hide or skin.

Colorado steer. A side-branded steerhide, not necessarily from Colorado.

Comber leather. A steer-hide leather, heavily stuffed and usually hand boarded, used in textile combing machines.

Combination tanned. Formerly, tanned with a blend of vegetable extracts. Today, tanned with two or more types of tanning materials such as chromium compounds and vegetable extracts, or chromium compounds and synthetic tanning materials.

Cordovan. Leather made from the tight firm shell portion of horse butts. Cordovan has very fine pores and a characteristic finish, and is very durable.

Corrected grain. (See Snuffed top grain.)

Country hides. Hides taken off by butchers and farmers. The quality is usually lower than that of packer hides because country hides are removed by less skilled hands and are not cured as well as packer hides.

Cowhide. Term specifically applied to leather made from hides of cows, although the term is sometimes loosely used to designate any leather tanned from hides of animals of the bovine species.

Crop. A side with the belly trimmed off. (OO'P'P in Figure 1.)

Crushed leather. Chrome-vegetable retanned leather with the grain accentuated by plating or other process.

Crust. Used as an adjective or in the phrase "in the crust." Refers to leather that has been tanned but not finished. (See Rough.)

Curing. Treating raw hides or skins so as to minimize putrefaction and bacterial action, but to enable the skins to be wet back conveniently in preparation for tanning. (See Brining, Dry salting, Dry pickling, Green salting and Pickle.)

Curling temperature. The temperature at which noticeable curling occurs, when gradually heating a leather specimen in water.

Currying. A process of treating tanned hides with oils and greases to prepare them for belting, sole, harness leathers, etc.

Deep buff. The first cut or split underneath the top grain or machine buff on which no traces of the grain remain.

Deerskin. In glove leather, a deerskin tanned and finished with the grain surface intact.

Degrained leather. Leather from which the grain has been removed after tanning, by splitting, abrading, or other process.

Degras, moellon. The direct oxidized oil pressed out of sheepskin after tannage with cod or other oil.

Doeskin. Commercial term for white leather from sheep or lambskin, tanned with alum and/or formaldehyde.

Double-dressed. As applied to chamois skins, with the grain removed and buffed or sueded on both surfaces.

Double shoulder. The fore part of the hide cut off at right angles to the backbone line at the break of the fore flank, with the belly cut off and the head cut off behind the horn hole. (R'UT'S' in Figure 1.)

Drawn grain. Shrunken, shriveled, or wrinkled grain surface of leather.

Drumhead leather. (See Parchment.)

Dry pickling. A method of curing skins from wool sheep with sodium sulfate and sodium chloride.

Dry salting. A method of curing hides in which the hides are first green-salted and then dried.

Dubbing (Also Dubbin). A mixture primarily of oils and fats used for restoring fatty matter to military footwear in the field.

Electrified shearling. Shearling in which the wool has been straightened by a special process. (Also Electrified lambskin.)

Elk leather. Trade term used to designate chrome-tanned cattlehide for uppers of work shoes, hunting boots, some children's shoes, and others requiring flexibility and durability. More properly, elk-finished cowhide. Leather from elk hide is more properly called "buckskin".

Embossed leather. Leather which has been ornamented with a geometrical or fancy design by heavy pressure in a machine.

Extract. A liquid, powder, or solid concentrate of vegetable tannin obtained by the extraction of tannin from natural sources.

Factory sole leather. One of the two principal types of sole leather. It is tanned and finished to have more flexibility and compressibility than finder's sole leather, and is more suitable for use in shoemaking machinery. (See Finder's sole leather.)

Fancy leather. Leathers made from hides and skins of all kinds which have commercial importance and value primarily because of grain or distinctive finish, whether natural or the result of processing. Such processing may be graining, printing, embossing, ornamenting (including gold, silver, and aluminum finishes), or any other finishing operation enhancing the appeal of the leather.

Fatliquor. An emulsion of oils or greases in water, usually with an emulsifying agent, used to lubricate the fibers of the leather.

Fat wrinkle. Wrinkles in the grain of leather, caused by fat deposits in the live animal.

Fiberboard. A firm, but somewhat flexible, composition material in sheet form, made from new, long vegetable fibers. Used for counters, insoles, midsoles, and heel lifts. The term is often loosely applied to boards made from scrap material or short-fibered stock, such as chip-board, which has inferior physical properties in the uses mentioned. (See also Leatherboard.)

Finder's sole leather. One of the two principal types of sole leather. It has less flexibility and compressibility than factory sole leather and is more suitable for use in shoe repair. (See also Factory sole leather.)

Finish. Materials applied to the grain and sometimes the split surface of the leather to cover blemishes, create smoothness, and give uniformity of color and appearance which may vary from dull to glossy.

Flesh. The inner side of a hide or skin.

Flesher. The flesh split or undercut of a sheepskin, split before tanning. (See Chamois.)

Flint dried. Dried in air without other curing.

Formaldehyde tannage. Tannage used especially for white leathers and washable glove leathers.

French kid. Leather tanned from kidskin by an alum or vegetable process.

Frigorifico hides. Cattlehides from South American slaughtering and freezing plants, cured in brine and salted.

Frizing. In tanning Mocha glove leather, a process of removing the grain surface involving long liming for not less than a month, during which the elastic structure of the grain layer is destroyed. (Also Friezing)

Front. The forepart of a hide or skin. Particularly in horsehide leathers, the front is used for garments, baseballs, etc. It is the part left when the butt is cut off about 22 inches from the root of the tail.

Full grain. Having the original grain surface of the skin.

Gasket leather. (See Hydraulic leather.)

Gill box leather. A leather used in textile machinery, similar to comber leather.

Glazed (Glace) kid. Chrome-tanned goatskin and kidskin leather, either black or in colors, which has a glazed finish.

Glove leather. Term covering two distinct classes: (1) The leather used for dress gloves (including those for street, riding, driving, and sports wear) made predominantly from sheep and lamb skins and to a lesser degree from deer, pig, goat, kid and Mocha skins. (2) The leather used for utilitarian or work gloves made from a variety of hides and skins of which the most important are horsehides, cattlehide splits, calfskins, sheepskins and pigskins.

Glove splits. Split chrome-tanned cattlehide leather used for work gloves.

Grain. The outer or hair side of a hide or skin. Also used as an adjective referring to that side.

Grained leather. Any leather on which the original natural grain has been changed or altered by any method, process or manipulation.

Green salting. A process of curing hides by treating them with salt on the flesh side and stacking in piles to cure for a period of ten days or more.

Grub hole. A hole through the hide caused by the penetration of the grub of the warble fly.

Gusset leather. A soft flexible leather used for gussets in shoes, bags, and cases.

Hair-on leather. Leather tanned without removing the hair from the skins or hide.

Hand buffs. A term used to describe upholstery leather of the same type as full top grain except that the surface of the hide is lightly snuffed or sandpapered all over. Such snuffing removes only the top of the hair follicles. (Also snuffed top grain, corrected top grain, top grain snuffed.)

Harness leather. A self-explanatory term sometimes so defined as to include collar and saddlery leathers. Harness leather, including the related items mentioned, is practically all made of vegetable-tanned cattlehides except for a considerable quantity of pigskins used for making saddle seats.

Hat leather. Usually sheepskin or calfskin for sweatbands of hats. The grain splits of sheepskin are vegetable-tanned for this purpose.

Head. That portion of the hide from the snout to the flare into the shoulder.

Heavy leather. A somewhat indefinite term, generally understood to include vegetable-tanned sole, belting, strap and mechanical leathers made from unsplit cattlehides. Also refers to the thick side of leather.

Hide. The pelt of a large animal, such as cow, horse, etc. Also used interchangeably with skin.

Hide grades. Standard hide grades, take-up and delivery practice are given in the booklet, "Approved Standard Practice Governing the Take-Up and Delivery of Domestic Packer Hides", published by the Tanners' Council of America, 411 5th Avenue, New York 16, New York.

Hide powder. Purified, shredded rawhide used as a reagent in the determination of tannins. Standard hide powder is any lot of hide powder officially approved by the American Leather Chemists Association.

Hide substance. The nitrogen content of leather multiplied by the factor 5.62.

Horsehide leather. Leather made from the hide of a horse or colt. (See Cordovan and Front.)

Hydraulic leather. A collective term sometimes used for the cattlehide leathers (vegetable, chrome, or combination tannage) with special stuffing added, which are used in pump valves, as piston packing, and so forth.

Indian tanned. Combination tanned with alum and vegetable tannins.

India-tanned. Term applied to hides and skins from India, considered as a semi-tanned raw material and generally retanned in the U.S.A. before finishing.

Iron. A term used for measuring thickness of sole leather. One iron equals 1/48 inch (0.53 millimeters).

Iron tannage. Tannage with salts of iron.

Kangaroo. Leather made from the hide of the kangaroo.

Kid. Originally referring to leathers made from the skins of immature goats, the term is now rather loosely applied to glove and shoe leathers made from goatskins.

Kip. Skin from a bovine animal in size between a calf and a cow, weighing in green-salted condition approximately from 15 to 30 pounds.

Lace leather. A form of rawhide leather (from cattlehides) for lacing sections of power-transmission belts; sometimes prepared also with an alum and oil, chrome, or combination tanning.

Lambskin leather. Term applied to leather from either lambskins or sheepskins, which are practically indistinguishable after tanning.

Larrigan leather. Oil-tanned light cattlehides, used largely for moccasins.

Latigo leather. A type of lace leather, alum and vegetable tanned, used in saddlery.

Leatherboard. A type of fiberboard in which the fiber content is at least 75 per cent leather, usually with asphaltic or resinous binder.

Levant. Term applied to goatskin on which the grain pattern is accentuated in tannage. Goatskin embossed to give a Levant pattern is properly described as "Levant-grained goatskin." Sheep, seal, and other skins bearing this pattern should not be described as "Levant leather" but as "Levant-grained sheepskin," etc.

Lining leather. Any leather used for making shoe linings, which includes sheep, lamb, kid, goat, cattle, calf, and splits.

Load. The amount of nonprotein material in vegetable-tanned leather.

Loading. The addition of glucose, magnesium sulfate, or other materials to give leather the physical properties needed for working in modern shoe machinery. (Also known as Filling or Stuffing.)

Machine buffs. That cut of the hide from which a buffing of approximately 1/64 inch (one ounce) in thickness has been removed from the grain. This should leave a portion of the grain on almost the entire hide.

Manufacturer's leather. (See Factory sole leather.)

Matadero hides. Hides from Argentina corresponding to city butcher or smaller packer hides of the United States.

Mechanical leather. A collective term for many types of leather used in connection with textile and other machinery.

Meter leather. A specialty leather made mainly from sheepskins treated to make it impermeable and used for the measuring bags of gas meters.

Mineral tanned. Tanned with chemical compounds of mineral origin, (chromium, zirconium or alum compounds) without the use of vegetable tanning materials.

Moellon. (See Degras.)

Morocco grain. Vegetable-tanned fancy goatskin leather having a distinctive pebbled grain.

Morocco leather. Vegetable-tanned fancy goatskin leather having a distinctive pebbled grain.

Mouton. A sheepskin shearling tanned and finished for use as a fur; usually with wool straightened.

Mukluk leather. Leather usually made from deer, elk and similar skins. It is tanned white with formaldehyde alum or syntans. It is very permeable to moisture vapor and retains its flexibility at low temperatures.

Napa leather. Chrome, alum, or combination tanned sheepskin glove leather, drum colored.

Native hide. A cattlehide without a brand.

Oak tannage. Originally, the tannage of leather entirely (or nearly so) with oak bark, later the tannage with a blend containing oak tannin. Now loosely applied to any tannage of heavy leather with vegetable extracts.

Offal. Parts of hides not used for standard grades of outsole leathers; the heads, shoulders, and bellies of heavy leather.

Oiling off. Coating the surface of leather with oil

Oil tannage. Tannage with cod oil or other oxidizing oil, usually of marine origin.

Ooze. Traditionally refers to a nap produced on vegetable-tanned leather. Also refers to other tannages sueded or napped on the grain side.

Ounce. A term used to indicate weight or substance of certain kinds of leather (such as upholstery, bag, and case leather). In theory it is based upon the assumption that one square foot of leather will weigh a certain number of ounces and will uniformly be of certain thickness; hence, a three-ounce leather theoretically would be one square foot of leather weighing three ounces. In practice, this varies because of specific gravity of tanning materials used and for that reason a splitter's gauge has been adopted which controls the commercial thickness of leather when sold by the square foot. An ounce is equivalent in thickness to $1/64$ inch = 0.0156 inch = 15.6 mils = 0.4 millimeters.

Pac leather. Highly water resistant leather used by lumber men, hunters and others for outdoor use.

Packer hides. Hides from meatpacking houses.

Packing leather. (See Hydraulic leather.)

Parchment. Traditionally alum-tanned sheepskin or slunk used for special documents, drum heads, lamps, etc.

Patent leather. Leather with a glossy impermeable finish produced by successive coats of drying oils, varnish, or synthetic resins.

Pebbled grain. An embossed-leather grain finish resembling a pebbled surface, ranging from fine pebbled Morocco goat to heavy Scotch grain upper leather.

Peccary. A wild boar found in Central and South America. The skin is usually chrome tanned and shaved to light weight for glove leathers. It is distinguishable from pigskin and Carpincho skins by the fact that bristle holes occur in straight line groups of three.

Pelt. A raw skin with the hair on. Usually refers to fur animals.

Persians. India-tanned hair sheepskins.

Picker leather. Leathers used for pickers in textile machinery, and having a wide range of properties. Some are hard rawhide buffalo leathers, others glycerine-treated rawhide and still others belting leather.

Pickle. To treat unhaired hides with a solution of salt and acid in order to prepare them for tannage or for temporary preservation until they reach the tannery.

Pigment-finished leather. Leathers finished with compounds containing opaque pigments which more or less conceal the grain pattern. Split leathers are often finished with pigments and embossed to simulate grain.

Pigskin. Leather made from the skins of pigs or hogs. In the glove leather trade "Pigskin" includes peccary and carpincho.

Pin Seal. Natural grain sealskin tanned for fancy leather. Imitations on other skins should be described as "pin-grain sheepskin", pin-grain goatskin", etc.

Pipeyness. Characteristic of loose grain leather which forms coarse wrinkles on bending with the grain inward.

Plating. Pressing leather with a heated metal plate, usually smooth, under high pressure.

Pocket-shaped. As applied to chamois skins, a skin trimmed in the form of a rectangle with the two corners at one end rounded.

Quebracho. A tanning material extracted from the wood of a South American tree.

Rawhide. Cattlehide that has been dehaired, limed, often stuffed with oil or grease, and has sometimes undergone other preparation, but has not been tanned. It is used principally for mechanical purposes, such as belt lacings, loom pickers, gaskets, pinions, gears, and for hand luggage, shoe laces, snow shoes, etc.

Raw streak. An untanned center layer of leather, visible in cross section as a light-colored streak, especially as applied to heavy leather.

Reconstituted leather. Material composed of collagen fibers, obtained from macerated hide pieces, which have been reconstructed into a fibrous mat.

Retan. A modifying secondary tannage applied after intermediate operations following the primary tannage.

Rigging leather. A strong flexible, vegetable-tanned leather.

Roan. A sheepskin, not split.

Roller leather. Vegetable-tanned sheep or calfskins used for cots or covers on the upper rolls of cotton-spinning machinery.

Rolling. A tannery operation in which the grain surface is compressed and smoothed under pressure by rollers.

Rough, rough tanned, in the rough. Terms applied to cattlehide leathers tanned but not finished. (See Crust.)

Russet. A term of varied meaning in the leather trade, since it connotes both color and tannage. Russet calf is the natural color of unfinished calf leather resulting from tannage by vegetable extracts. Russet harness is a completely finished leather of bright, clean, uniform color and finish. Russet sheepskin originally was leather tanned in cold-leached hemlock bark, used for shoe linings, with color resulting from the hemlock; now sheepskin colored as though tanned by vegetable extracts. Russet upholstery is leather tanned but not finished.

Russia leather. Originally a Russian calfskin shoe leather distinguished by its odor of birch oil. Now in the U.S.A., a fancy leather.

Saddle leather. Vegetable-tanned cattlehide leather for harness and saddlery, usually of a natural tan shade and rather flexible.

Saladero hides. Argentinian hides corresponding to small-packer hides in the U.S.A.

Salt stain. Discoloration on the surface of hides and skins, developed during the curing process.

Scotch grain. A pebbled pattern embossed on cattlehide or calf leather.

Scud. Remnants of epithelial tissue, hair, dirt, etc., left in the hair follicles after unhairing.

Scudding. Removal of scud from unhaired hides by scraping with a blade, either by hand or machine.

Shank. Leg portion of hide pattern. (See Figure 1.)

Sharkskin. Leather made from the top grain of the skins of sharks. It has various natural grain markings. The term should not be applied to leather made from other skins and embossed.

Shearling. Leather made from sheepskin with the wool on. This wool is sheared to the desired length shortly before slaughter. The short wool is left on the skin when tanned.

Shell. A portion from the butt end of a horsehide, from which leather is of tight, firm fiber structure. (See Cordovan.)

Shoulder. Half of a double shoulder. The fore part of the hide, cut off at right angles to the back bone line at the break of the fore flank, with the belly cut off and the head cut off behind the horn holes. (R'UZY in Figure 1.)

Shrinkage temperature. The temperature at which measurable shrinkage occurs when leather is gradually heated in air or in a fluid (usually water). Wet shrinkage is called hydrothermal.

Side. A side is half a hide cut along the back bone line with the tail not more than 6 inches long. (OO'WV in Figure 1.)

Side leather. Shoe upper leather made from the grain side of cattlehides. The name comes from the practice of cutting the hide along the backbone into two sides before tanning. The skins are usually shaved on the flesh side to uniform thickness and the grain is corrected.

Skin. The pelt of a small animal, such as calf, pig, sheep, etc. Also used interchangeably with "hide."

Skiver. The grain split of a sheepskin used for hat sweatbands and small leather goods.

Skiving. Cutting off a thin layer of leather to bring it to uniform thickness.

Slab. (1) See Split. (2) In belting leather, the parts of a bend left after the centers are cut out.

Slack tannage. (1) Incomplete tannage. (2) Light tannage, deliberately less than usual.

Slats. Dried, untanned sheepskins with little or no wool.

Slunk. The skin of an unborn or prematurely born animal, especially calf.

Snuffed top grain (top grain snuffed). Portions of the grain surface lightly abraded with emery wheel or sandpaper, so as to lessen the effect of grain damage.

Sole leather butt bend. A double bend. (PR'S'T in Figure 1.)

Spew. Any constituents of leather that come to the surface in the form of a white crystalized deposit or a dark gummy deposit. (Also Spue)

Split. A term used to describe the portion of hide or skin, split into two or more thicknesses, other than the grain or hair side. Splits are usually named according to their sequence of production, such as "main," "second," or "slab" split (in case of upholstery leather); or for the use to which they are to be put, such as "flexible" (for innersoles), "glove," "waxed" (for cheap shoe-uppers); "bag and case" (finished with pyroxylin or pigment finish), "sole," etc.

Splitting. (1) Cutting leather into two or more layers. (See also Upholstery leather) (2) Cutting a hide into two sides preparatory to tanning.

Spready hide. A hide of large area in proportion to the weight.

Steerhide. (See Hide grades.)

Strap bellies. Thin, light-weight, vegetable-tanned cattlehide bellies, rather flexible and with low load, processed for the strap trade.

Stuffing. The process of incorporating grease in leather by drumming the wet leather with warm molten grease and oils.

Sulfite cellulose. A by-product of paper mills, produced in sulfiting wood pulp, used as a tanning material more correctly named lignosulfonate since it does not contain cellulose.

Syntan. A synthetic organic tanning material.

Table dyeing. The application of dyestuff to leather with a brush, the leather being laid on a table. (Also called Brush coloring.)

Table run. Used to describe leather that has not been sorted and graded before selling by the tanner. (Also Tannery run, or T.R.)

Tannery run. (See Table run.)

Tawing. The old English term applied to the process of making leather with alum as distinguished from tanning which was originally confined to vegetable tanning.

Top grain. The grain side of a hide from which nothing except the hair and associated epidermis have been removed by reduction to a specific thickness by shaving, splitting, or other means.

Trim. The removal of parts of a raw hide not suitable for making leather, such as portions from the outer edges of heads, shanks and bellies.

Upholstery leather. A general term for leather processed for use for furniture, airplanes, busses, and automobiles. The staple raw material in this country consists of spready cattlehides, split at least once and in many cases two or three times. The top or grain cuts go into the higher grades and the splits into the lower grades.

Valve leather. (See Hydraulic leather.)

Vat dyeing. The application of dyestuffs to leather by the immersion of the leather in a revolving drum containing the dyestuff solution, as contrasted with "Table dyeing".

Veal. A large calfskin, almost as large as kip.

Vegetable tanning. The conversion of rawhides into leather by treating with water solution of tannin extracted from materials of vegetable origin.

Veiny. Appearance of leather characterized by many clearly visible blood vessels mostly on the flesh side, either closed or cut open by buffing or shaving operation.

Vellum. (See Parchment.)

Wallaby. Leather from skin of the wallaby, a small or medium-sized species of kangaroo.

Walrus. Leather from the hides of walrus. Walrus hide is very thick and is used for buffing wheels. When split, it is used for bag leather. Split walrus and seal leather are practically indistinguishable, and "walrus leather" in the traveling-goods industry is used to refer to sealskin leather on which a simulated walrus grain is embossed.

Welting shoulder. The shoulder portion of vegetable-tanned cattlehide leather, tanned with a low load to give the flexibility required for a welt.

White weight.   The weight of limed and unwashed stock.

Willow.   (1) Willow grain—refers to boarded leather. (2) In the sporting goods industry, Willow tanned is used to indicate flexible, well-oiled, chrome-tanned cattlehide or horsehide used for gloves.

Window.   In a chamois skin, a thin portion that transmits light when the skin is viewed against a window or light background.

Woolskin.   Sheepskin with the wool on.

Wrinkle.   A permanent crease or furrow on the grain surface of a hide or leather, incapable of removal by rolling or plating.

INDEX